NB 7254636 0

D1406201

# Materials and their Uses

# Materials and their Uses

W. Bolton

BUTTERWORTH
HEINEMANN

Butterworth-Heinemann Ltd
Linacre House, Jordan Hill, Oxford OX2 8DP

 A member of the Reed Elsevier plc group

OXFORD LONDON BOSTON
NEW DELHI SINGAPORE SYDNEY
TOKYO TORONTO WELLINGTON

First published 1996

© W. Bolton 1996

**British Library Cataloguing in Publication Data**
Bolton, W.
  1 Materials and their Uses
  I. Title


ISBN 07506 2726 3

**Library of Congress Cataloging in Publication Data**
A catalogue record for this book is available from the Library of Congress




Printed and bound in Great Britain by
Martins the Printers Ltd, Berwick upon Tweed

# Contents

# Preface

This book has been written to provide a comprehensive coverage of the mandatory unit Investigate Materials and their Uses for the Advanced GNVQ in Science, and also cover the materials modules in A-level Physics syllabuses, e.g. Module PH4A in the University of London syllabus, and A-level Design and Technology syllabuses. It aims to introduce science students to how scientists determine the properties of materials and evaluate their suitability for different uses, how the properties of materials can be changed to make them more useful and how the outcome of such work is of importance to engineers, designers, technologists, builders, etc. No prior knowledge is assumed but it is expected that students will have a basic knowledge of science and mathematics. The book is also seen as being of relevance to other courses where a knowledge of materials is required.

The aims of the chapters and their relationship to the elements of the GNVQ unit are:

| Chapter | Aim | GNVQ element |
|---|---|---|
| 1 | Identify the purposes for which materials are needed and the conditions under which they are used. | 2.1 |
| 2 | Describe the characteristic properties of materials: mechanical, electrical, thermal, optical, and chemical. | 2.1 |
| 3 | Describe how the physical properties of materials can be determined. | 2.2 |
| 4 | Describe the structure of solids: metals, polymers and ceramics. | 2.2 |
| 5 | Describe how the structure of materials is related to properties and how materials can be modified to make them more useful. | 2.2 2.3 |
| 6 | Justify the selection of material for particular uses. | 2.1 2.2 2.3 |

There are worked examples in the text. In addition, at the ends of each chapter, there are multiple-choice questions and problems. Answers are given for all the multiple-choice questions and guidance given as to the answers for all the problems.

W. Bolton

# 1 Materials and their uses

## 1.1 Purposes for which materials are used

In considering materials and their uses we need to consider the function or functions required of the product being made and hence the conditions under which the materials will be used and the properties required of them. This chapter is an introduction to a general consideration of the functions required of materials. Chapter 2 extends this with a specification of the properties of materials, Chapter 3 indicating how such properties can be measured.

Consider a product – a drinking vessel. It has the primary function of holding a liquid. Such products might take the form of a pottery cup, a plastic cup, a glass, a pewter tankard, an expanded polystyrene beaker, etc. A variety of materials are used and are able to give a product which fulfils the primary function. If we restrict the function to a drinking vessel capable of holding a hot liquid, then we might add the condition that the drinking vessel can be picked up by a human hand without burning or causing discomfort. We might solve this problem by adding a handle with the result that the cup is held at a point remote from the hot liquid. This would mean that we could use a material which is a poor but not excessively bad conductor of heat, e.g. a pottery cup. Alternatively we might choose a material for the cup which is a very bad conductor of heat and so does not need a handle, e.g. a polystyrene beaker.

It is necessary to distinguish between the functions required of the product and those required of the material. We might be able to meet a particular requirement by product design methods, as for example by designing a handle for a cup for hot liquids. Alternatively we might adopt a materials solution to the problem, as by the use of polystyrene as the material for a beaker. This book is primarily concerned with materials solutions to problems.

### Example

Pewter, a metal, is used for drinking tankards. Since metals are good conductors of heat, what does this say about the liquids that the tankard is designed to hold?

Assuming that there is no handle made of another material, then the tankard would have been designed to hold cold liquids since hot liquids would make it uncomfortable to hold.

## 1.2 The requirements of materials

What materials could be used for Coca-Cola containers? Well you can buy Coca-Cola in aluminium cans, in glass bottles and in plastic bottles. What makes these materials suitable and others not? In order to attempt to answer this question we need to discuss the purposes for which the materials are needed. The primary function of the container is to be able to hold a liquid. We might then talk about the need for the container material to be:

1  Rigid, so that it does not stretch unduly, i.e. become floppy, under the weight of the Coca-Cola.
2  Strong, so that it can stand the weight of the Coca-Cola without breaking.
3  Resistant to chemical attack by the Coca-Cola.
4  Able to keep the 'fizz' in the Coca-Cola, i.e. not to allow the gas to escape through the walls of the container.
5  Low density so that it is not too heavy.
6  Cheap.
7  Easy and cheap to process to produce the required shape.

You can no doubt think of more requirements. You might consider that the container, should be capable of being recycled and so reduce the demand on the earth's resources. The selection of a material involves balancing a number of different requirements and making a choice of the material which fulfils as many requirements as possible, as well as possible.

Consider another product, a bridge. What are the requirements for the material to be used for the beams in a bridge? These are likely to include:

1  Strength, so that when the bridge is subject to loads crossing it such as people, cars, lorries, etc., it will not break.
2  Stiff enough so that the bridge will not stretch unduly under the load.
3  Can be produced and joined in long enough lengths to be able to span the gap to be bridged.
4  The materials and fabrication costs are not too high.
5  Resists or can be protected from atmospheric corrosion.
6  Can be maintained at a reasonable cost over a period of years.

You can no doubt add more requirements. Materials that are used include wood, steel and reinforced concrete.

Consider another example, the requirements for the materials used for the conductors in an electric cable. These are likely to include:

1  A very good conductor of electricity.
2  Flexible so that cables can easily be bent round corners.
3  Can be produced with a circular cross-section in long lengths with small diameters.
4  Cheap.

The most commonly used material is copper.

Consider another product, a surgical implant such as the part designed to replace the bone structure of the hip-joint (Figure 1.1). A hip-joint is subject to static loading as a result of the muscle action which keeps the parts of the joint together and repeated loading caused by walking and other activities. Such a surgical implant must therefore be of a material which has the following properties:

1  Be biocompatible, i.e. be accepted by the body as though it were its own bone.
2  Have high corrosion resistance. This is because body fluids are aqueous salt solutions, not too dissimilar to sea water, and so can cause corrosion of many metallic materials.
3  Have high wear resistance since one part of the joint is repeatedly rubbing against the other.
4  Have sufficient strength and toughness to withstand the loading they are subject to.
5  Be able to withstand the repeated loading caused by walking, etc. The material is said to require adequate fatigue strength.

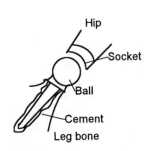

Figure 1.1 *Hip-joint implant*

Stainless steel or a titanium alloy for a carbon-fibre composite have been used for the ball part and a polymer for the socket.

The selection of a material depends on the properties required of it in order that it can fulfil the required uses. Chapter 2 gives definitions of commonly met properties.

### 1.2.1 Factors determining the selection of materials

The selection of a material from which a product can be manufactured depends on a number of factors. These are:

1  The requirements imposed by the conditions under which the product is used, i.e. the service requirements. Thus if a product is to be subject to forces, then it might need strength; if subject to a corrosive environment, then it might require corrosive resistance.
2  The requirements imposed by the methods proposed for manufacturing the product. For example, if a material has to be bent as part of its processing, the material must be ductile enough to be bent without breaking. A brittle material could not be used.
3  The requirements imposed by environmental considerations. For example, should be material used be selected because it is capable of being recycled or does not result in pollution?
4  Availability and cost.

### Example

What functions might be required of a material for the rainwater gutters and drainpipes used with houses.

The service requirements might be considered to be for a rigid, durable material that is capable of withstanding an outdoor environment without deteriorating. It should be capable of being processed in reasonable long lengths and be cheap. A material that is used is the polymer unplasticised PVC.

## 1.3 The range of materials

Materials are usually classified into four main groups:

1   *Metals*
In general, metals have high electrical conductivities, high thermal conductivities, can be ductile and thus permit products to be made by being bent into shape, and have a relatively high stiffness and strength. Engineering metals are generally alloys. The term *alloy* is used for metallic materials formed by mixing two or more elements. For example, mild steel is an alloy of iron and carbon, stainless steel is an alloy of iron, chromium, carbon, manganese and possibly other elements. The reason for adding elements to the iron is to improve the properties. Pure metals are very weak materials. The carbon improves the strength of the iron. The presence of the chromium in the stainless steel improves the corrosion resistance.

2   *Polymers and elastomers*
Polymers can be classified as either *thermoplastics* or *thermosets*. Thermoplastics soften when heated and become hard again when the heat is removed. The term implies that the material becomes 'plastic' when heat is applied. Thermosets do not soften when heated, but char and decompose. Thus thermoplastic materials can be heated and bent to form required shapes, thermosets cannot. Thermoplastic materials are generally flexible and relatively soft. Polythene is an example of a thermoplastic, being widely used in the forms of films or sheets for such items as bags, 'squeezy' bottles, and wire and cable insulation. Thermosets are rigid and hard. Phenol formaldehyde, known as Bakelite, is a thermoset. It is widely used for electrical plug casings, door knobs and handles. The term *elastomers* is used for polymers which by their structure allow considerable extensions that are reversible. The material used to make rubber bands is an obvious example of such a material.

All thermoplastics, thermosets and elastomers have low electrical conductivity and low thermal conductivity, hence their use for electrical and thermal insulation. Compared with metals, they have lower densities, expand more when there is a change in temperature, are generally more corrosion resistant, have a lower stiffness, stretch more and are not as hard. When loaded they tend to creep, i.e. the extension gradually changes with time. Their properties depend very much on the temperature so that a polymer which may be tough and flexible at room temperature may be brittle at 0°C and show considerable creep at 100°C.

3   *Ceramics and glasses*
Ceramics and glasses tend to be brittle, relatively stiff, stronger in compression than tension, hard, chemically inert and bad conductors of electricity and heat. Glass is just a particular form of ceramic, with

ceramics being crystalline and glasses non-crystalline. Examples of ceramics and glasses abound in the home in the form of cups, plates and glasses. Alumina, silicon carbide, cement and concrete are examples of ceramics. Because of their hardness and abrasion resistance, ceramics are widely used as the cutting edges of tools.

4 *Composites*

Composites are materials composed of two different materials bonded together. For example, there are composites involving glass fibres or particles in polymers, ceramic particles in metals (referred to as cermets), and steel rods in concrete (referred to a reinforced concrete). Wood is a natural composite consisting of tubes of cellulose in a polymer called lignin. Composites made with fibres embedded all aligned in the same direction in some matrix will have properties in that direction markedly different from properties in other directions. Composites can be designed to combine the good properties of different types of materials while avoiding some of their drawbacks.

Chapter 4 continues the above discussion with a consideration of the internal structure of materials.

**Example**

A material with high electrical conductivity is required. What group of materials is likely to be considered?

Metals have high electrical conductivities compared with polymers, ceramics or composites and thus a metal is likely to be used.

**Problems**

*Questions 1 to 11 have four answer options: A, B, C and D. Choose the correct answer from the answer options.*

Questions 1 to 3 relate to the following information:

The following are a number of purposes for which materials might be needed:

A   Protection from environmental damage.
B   Act as a supporting member.
C   Conduct electricity.
D   Conduct heat.

Select the most likely purpose from the above list of options for which the material to be used in each of the following instances is selected:

1   The connecting wires to a resistor.
2   The leg of a chair.
3   The glass envelope of an electric light bulb.

Questions 4 to 6 relate to the following information:

The following are some of the properties that materials might have:

A   Strength.
B   Density.
C   Brittleness.
D   Hardness.

Select the property that would indicate the suitability of a material for:

4   Steel used for load-bearing structures.
5   Steel used for the point of a drill bit.
6   Plastic used for a washing up bowl.

Questions 7 to 9 relate to the following information:

The following are some of the properties that materials might have:

A   Conductivity of heat.
B   Conductivity of electricity.
C   Transparency.
D   Hardness.

Select the property that would indicate the suitability of a material for:

7   A hacksaw blade.
8   The casing for an electrical plug.
9   The lens material for a vehicle rear light.

10  Decide whether each of the following statements is TRUE (T) or FALSE (F).

In general, metals are:
(i)   Good conductors of heat.
(ii)  Good conductors of electricity.

A   (i) T   (ii) T
B   (i) T   (ii) F
C   (i) F   (ii) T
D   (i) F   (ii) F

11  Decide whether each of the following statements is TRUE (T) or FALSE (F).

In general, ceramics are:
(i)   Good conductors of heat.
(ii)  Strong in compression and weak in tension.

A   (i) T   (ii) T
B   (i) T   (ii) F
C   (i) F   (ii) T
D   (i) F   (ii) F

**12** List the requirements of the materials for the following purposes:
(a) Table legs.
(b) A mirror for a large telescope.
(c) A spring.
(d) The wall of a pottery kiln.
(e) The forks of a bicycle.

# 2 Properties of materials

**2.1 Properties**   The selection of a material for a product is based upon a consideration of the properties required. These include:

1   *Mechanical properties.* These are the properties displayed when a force is applied to a material and include strength, stiffness, hardness, ductility toughness and wear resistance. The density property is also frequently considered in relation to the strength and stiffness.
2   *Electrical properties.* These are the properties displayed when the material is used in electrical circuits or components and include resistivity, conductivity and resistance to electrical breakdown.
3   *Thermal properties.* These are the properties displayed when there is a heat input to a material and include expansivity, heat capacity, thermal conductivity and melting point.
4   *Optical properties.* These are the properties involved when light passes through the material. They include such properties as the refractive index and transmissivity.
5   *Chemical properties.* These are, for example, relevant in considerations of corrosion and solvent resistance.
6   *Magnetic properties.* These are relevant when a material is considered as a magnet or part of an electrical component, such as an inductor which relies on such properties.

In this chapter the key properties listed above are discussed and the quantities that are used as a measure of them are defined.

In discussing the properties of materials it is important to recognise that they are often markedly changed by the temperature at which a material is being used and any treatments the materials undergo. For example, a plastic may be relatively stiff at room temperature but far from stiff at the boiling point of water. A steel may be ductile at 20°C but become brittle at temperatures below −10°C. Steels can have their properties changed by heat treatment, such as annealing which involves heating to some temperature and slowly cooling. This renders the material soft and ductile. Heating a steel to some temperature and then quenching, i.e. immersing the hot material in cold water, can be used to make a steel harder, stronger and less ductile. Materials can also have their properties changed by working. For example, if you take a piece of carbon steel and permanently deform it, by perhaps bending it, then it will have different mechanical properties to those existing before that deformation (see Chapter 5 for more information). It is said to be work-hardened.

## 2.2 Mechanical properties

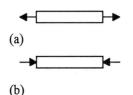

(a)

(b)

Figure 2.1 *(a) Tension, (b) compression*

Mechanical properties are about how materials behave when subject to forces. When a material is subject to external forces, then internal forces are set up in the material which oppose the external forces. The material can be considered to be like a spring. When it is stretched by external forces, it sets up internal opposing forces which are readily apparent when the spring is released and they force it to contract. When a material is subject to external forces which stretch it, then it is said to be in *tension* (Figure 2.1(a)). When a material is subject to forces which squeeze, it then is said to be in *compression* (Figure 2.1(b)). An object, in some situations, can be subject to both tension and compression. Consider a beam (Figure 2.2) which is being bent by forces. The bending causes the upper surface to contract and so be in compression and the lower surface to extend and so be in tension.

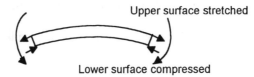

Figure 2.2 *Bending*

### 2.2.1 Stress and strain

In discussing the application of forces to materials, an important aspect is often not so much the size of the force itself as the size of the force applied per unit area. Thus, if we stretch a strip of material by a force $F$ applied over its cross-sectional area $A$, then the force applied per unit area is $F/A$ (Figure 2.3). The term *stress* is used for the force per unit area:

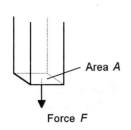

Figure 2.3 *Stress*

$$\text{stress} = \frac{\text{force}}{\text{area}}$$

Stress has the units of pascal (Pa), with 1 Pa being a force of 1 newton per square metre, i.e. 1 Pa = 1 N/m². Stresses are often rather large and so prefixes are used with the unit. Thus 1 kPa (kilopascal) = 1000 Pa, 1 MPa (megapascal) = 1 000 000 Pa. The area used in calculations of the stress is generally the original area that existed before the application of the forces, not the area after the force has been applied. This stress is thus sometimes referred to as the *engineering stress*, the term *true stress* being used for the force divided by the actual area existing in the stressed state.

When a material is subject to tensile or compressive forces, it changes in length. The term *strain* (Figure 2.4) is used for:

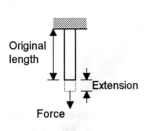

Figure 2.4 *Strain*

$$\text{strain} = \frac{\text{change in length}}{\text{original length}}$$

Since strain is a ratio of two lengths it has no units Thus we might, for example, have a strain of 0.01. This would indicate that the change in length is 0.01 × the original length. However, strain is frequently expressed as a percentage.

$$\text{Strain as a \%} = \frac{\text{change in length}}{\text{original length}} \times 100$$

Thus the strain of 0.01 as a percentage is 1%, i.e. the change in length is 1% of the original length.

### Example

A bar of material with a cross-sectional area of 50 mm$^2$ is subject to tensile forces of 100 N. What is the tensile stress?

The tensile stress is the force divided by the area and is thus:

$$\text{tensile stress} = \frac{100}{50} \text{ N/mm}^2$$

To put the stress in units of pascals, we need to have the area in square metres. Thus:

$$\text{tensile stress} = \frac{100}{50 \times 10^{-6}} = 2 \times 10^6 \text{ N/m}^2 \text{ or Pa}$$

We can write this stress in terms of megapascals as 2 MPa.

### Example

A strip of material has a length of 50 mm. When it is subject to tensile forces it increases in length by 0.020 mm. What is the strain?

The strain is the change in length divided by the original length and is thus:

$$\text{strain} = \frac{0.020}{50} = 0.000\,4$$

Expressed as a percentage, the strain is:

$$\text{strain} = \frac{0.020}{50} \times 100 = 0.04\%$$

### Example

What is the stress experienced by a leg bone of cross-sectional area 9.5 cm$^2$ when it experiences a compressive force of 850 N during walking?

Since 1 cm$^2$ = 10$^{-4}$ m$^2$, the area is 9.5 × 10$^{-4}$ m$^2$. Hence, since stress is force/area

$$\text{stress} = \frac{850}{9.5 \times 10^{-4}} = 895 \text{ kPa}$$

### 2.2.2 Stress–strain graphs

If gradually increasing tensile forces are applied to, say, a length of steel wire or a strip of mild steel, then initially when the forces are released the material springs back to its original shape. The material is said to be *elastic*. If measurements are made of the extension at different forces and a graph plotted, then the force needed to produce a given extension is found to be proportional to the extension and the material is said to obey *Hooke's law*.

force ∝ extension

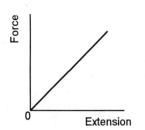

Figure 2.5  *Hooke's law*

Figure 2.5 shows a graph when Hooke's law is obeyed. Such a graph applies to only one particular length and cross-sectional area of a particular material. We can make the graph more general so that it can be applied to other lengths and cross-sectional areas of the material by dividing the extension by the original length to give the strain and the force by the cross-sectional area to give the stress. Then we have, for a material that obeys Hooke's law,:

stress ∝ strain

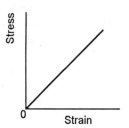

Figure 2.6  *Stress–strain graph when Hooke's law applies*

The stress–strain graph (Figure 2.6) is just a scaled version of Figure 2.5.

However, if the forces are made large enough, materials generally do not keep on obeying Hooke's law. The point at which this occurs is called the *limit of proportionality*. When a particular level of force is reached the material stops springing back completely to its original shape, i.e. being *elastic*, and is then said to show some *plastic* behaviour. The term plastic is used for that part of the behaviour in which permanent deformation occurs. This point often coincides with the point on a force–extension, or stress–strain, graph at which the graph stops being a straight line graph, i.e. the *limit of proportionality*. The stress at which the material starts to behave in a non-elastic manner is called the *elastic limit*.

Figure 2.7 shows the type of stress–strain graph which would be given by a sample of mild steel. Up to the limit of proportionality, Hooke's law is obeyed and the material shows elastic behaviour; beyond it, there is a mixture of elastic and plastic behaviour. Generally at about this stress, the material begins to stretch without any further increase in force and is said to have yielded. This is the *yield point* on the graph. The term *yield stress* is used for the stress at which this occurs. With mild steel, the graph then shows a dip before an increase in extension requires an increase in force again. The term *tensile strength* is used for the maximum value of the stress that the material can withstand without breaking. The graph may show a

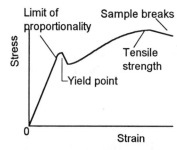

Figure 2.7  *Stress–strain graph for mild steel*

slight decline beyond this point before the mild steel sample breaks. The *tensile strength* is defined as the maximum tensile stress the material can withstand without breaking, i.e.:

$$\text{tensile strength} = \frac{\text{maximum tensile forces}}{\text{original cross-sectional area}}$$

The *compressive strength* is the maximum compressive stress the material can withstand without becoming crushed. The unit of strength is the pascal (Pa), with 1 Pa being 1 $N/m^2$. Strengths are often millions of pascals and so the unit MPa is often used, 1 MPa being $10^6$ Pa or 1 000 000 Pa.

Figure 2.8 shows the stress–strain graphs for a number of materials. The stress–strain graph for cast iron (Figure 2.8(a)) is virtually just a straight-line graph with virtually all elastic behaviour and little plastic deformation. The slight curved part at the top of the graph indicates a small departure from straight-line behaviour and a small amount of plastic behaviour. The graph indicates that the limit of proportionality is about 280 MPa and the tensile strength is about 300 MPa. The stress–strain graph for glass (Figure 2.8(b)) has a similar shape to that for cast iron, with virtually all elastic behaviour and little plastic deformation. The graph indicates that the limit of proportionality is about 250 MPa and the tensile strength about 260 MPa. The stress–strain graph for mild steel (Figure 2.8(c)) shows a straight-line portion followed by a considerable amount of plastic deformation. Much higher strains are possible than with cast iron or glass, i.e. mild steel stretches much more. The limit or proportionality is about 240 MPa and the tensile strength is about 400 MPa. The stress–strain graph for polyethylene (Figure 2.8(d)) shows only a small region where elastic behaviour occurs and a very large amount of plastic deformation possible. Very large strains are possible, a length of such material being capable of being stretched to almost four times its initial length. The limit of proportionality is about 8 MPa and the tensile strength is about 11 MPa. Table 2.1 in the next section contains typical values of yield stress and tensile strength for a range of materials.

The stretching properties of materials often change with time. Thus, for example, when a wire is stretched, it may show a tendency to elongate gradually for some time after the load was applied. This phenomena is termed *creep* and it is particularly significant with plastics.

### Example

A material has a yield stress of 200 MPa. What tensile forces will be needed to cause yielding with a bar of the material with a cross-sectional area of 100 $mm^2$ ?

Since stress is force/area, then:

$$\text{yield force} = \text{yield stress} \times \text{area} = 200 \times 10^6 \times 100 \times 10^{-6}$$

The yield stress is thus 20 000 N or 20 kN.

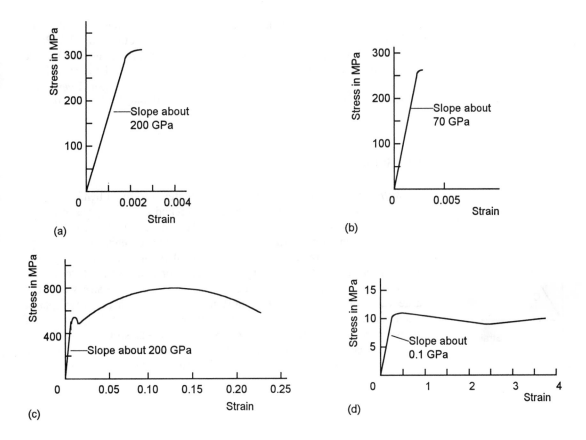

Figure 2.8 *Stress–strain graphs: (a) cast iron, (b) glass, (c) mild steel, (d) polyethylene*

### Example

A sample of an aluminium alloy has a tensile strength of 140 MPa. What will be the maximum force that can be withstood by a rod of that alloy with a cross-sectional area of 1 cm²?

Since 1 cm² = $10^{-4}$ m², the cross-sectional area of the rod is $1 \times 10^{-4}$ m². Hence, since the tensile strength is maximum force/area :

$$140 \times 10^6 = \frac{\text{maximum force}}{1 \times 10^{-4}}$$

and so maximum force = $140 \times 10^6 \times 1 \times 10^{-4} = 140 \times 10^2 = 14.0$ kN.

### Example

Given the polyethylene stress–strain graph shown in Figure 2.8(d), explain how it would feel if you pulled a strip of the material between your hands.

Initially the material would stretch very little as the forces applied by the hands increased. However, when the stress is about 8 to 10 MPa, it begins to stretch very easily and no increase in force is needed to give considerable extensions. If released, during this part of the stretching, the material would not spring back to its original length. Eventually, when the material is almost four times its initial length, the polyethylene breaks.

### 2.2.3 Stiffness

The *stiffness* of a material is the ability of a material to resist bending. When a strip of material is bent, one surface is stretched and the opposite face is compressed, as illustrated in Figure 2.2. The more a material bends, the greater is the amount by which the stretched surface extends and the compressed surface contracts. Thus a stiff material would be one that gave a small change in length when subject to tensile or compressive forces. This means a small strain when subject to tensile or compressive stress and so a large value of stress/strain. For most materials, a graph of stress against strain initially gives a straight-line relationship. Thus a large value of stress/strain means a steep initial gradient of the stress–strain graph (Figure 2.9). The quantity stress/strain when we are concerned with the straight line part of the stress–strain graph is called the *modulus of elasticity* (or *Young's modulus*), symbol $E$:

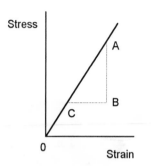

Figure 2.9 *Gradient is AB/BC*

$$\text{modulus of elasticity } E = \frac{\text{stress}}{\text{strain}}$$

The units of the modulus are the same as those of stress, since strain has no units. Engineering materials frequently have a modulus of elasticity of the order of 1 000 000 000 Pa, i.e. $10^9$ Pa. This is generally expressed as GPa, with 1 GPa = $10^9$ Pa. Typical values are about 200 GPa for steels and about 70 GPa for aluminium alloys. A stiff material has a high modulus of elasticity, thus steels are stiffer than aluminium alloys. For most engineering materials, the modulus of elasticity is the same in tension as in compression.

Table 2.1 gives typical values of yield stress, strength and modulus of elasticity for a range of materials.

### Example

For a material with a tensile modulus of elasticity of 200 GPa, what strain will be produced by a stress of 4 MPa?

Provided the stress does not exceed the limit of proportionality, since the modulus of elasticity is stress/strain

$$\text{strain} = \frac{\text{stress}}{\text{modulus}} = \frac{4 \times 10^6}{200 \times 10^9} = 0.000\,02$$

Expressed as a percentage, the strain is 0.002%.

Table 2.1 *Typical mechanical properties of materials*

| Material | Yield stress MPa | Tensile strength MPa | Modulus of elasticity GPa |
|---|---|---|---|
| Aluminium alloys | 50–300 | 100–400 | 70 |
| Bone: tensile | | 80 | 16 |
|    : compressive* | | 50 | 10 |
| Cast iron | 150–600 | 150–600 | 120–170 |
| Copper alloys | 60–300 | 160–600 | 70–150 |
| Glass | 30–90 | 30–90 | 70–80 |
| Mild steel | 230 | 400 | 210 |
| Polyethylene | | 30 | 1 |
| Polyvinyl chloride (PVC) | | 35–60 | 2–4 |
| Stainless steel | 200–1300 | 400–1800 | 210 |
| Wood (ash) | | 100 | 16 |

* All the values, with these exceptions, relate to the tensile properties of the materials.

**Example**

Which of the following plastics is the stiffest?

| | |
|---|---|
| ABS | tensile modulus 2.5 GPa |
| Polycarbonate | tensile modulus 2.8 GPa |
| Polypropylene | tensile modulus 1.3 GPa |
| PVC | tensile modulus 3.1 GPa |

The stiffest plastic is the one with the highest tensile modulus and so is the PVC.

**Example**

Bone has a compressive modulus of elasticity of about 10 GPa. If a leg bone with a cross-sectional area of 9.5 cm² experiences a force of 850 N during walking, what will be the percentage change in length by which the bone is compressed? Assume the limit of proportionality is not exceeded.

The percentage change in length is the strain expressed as a percentage. Since 1 cm² = $10^{-4}$ m², the area is $9.5 \times 10^{-4}$ m². Hence, since stress is force/area and the modulus of elasticity = stress/strain

$$10 \times 10^9 = \frac{\text{stress}}{\text{strain}} = \frac{\left(\dfrac{850}{9.5 \times 10^{-4}}\right)}{\text{strain}}$$

Hence

$$\text{strain} = \frac{\left(\dfrac{850}{9.5 \times 10^{-4}}\right)}{10 \times 10^9} = 8.95 \times 10^{-5}$$

Expressed as a percentage, the strain is $8.95 \times 10^{-3}\%$.

### 2.2.4 Energy stored

Consider a length of material being stretched by tensile forces. When a length of material is stretched by an amount $x_1$ as a result of a constant force $F_1$ then the work done is:

work = force × distance moved by point of application of force

and thus:

$$\text{work} = F_1 x_1$$

Thus if a force–extension graph is considered (Figure 2.10), the work done, when we consider a very small extension, is the area of that strip under the graph. The total work done in stretching a material to an extension $x$, i.e. through an extension which we can consider to be made up of a number of small extensions with $x = x_1 + x_2 + x_3 + ...$, is thus

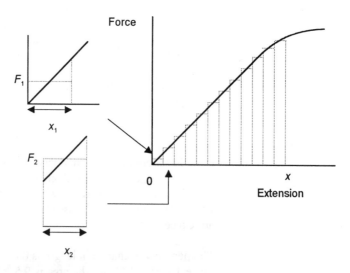

Figure 2.8 *Work done in extending a material*

$$\text{work} = F_1 x_1 + F_2 x_2 + F_3 x_3 + \dots$$

and so is the area under the graph up to $x$. If we divide both sides of this equation by the volume, i.e. the product of the cross-sectional area $A$ of the strip and its length $L$, we have:

$$\frac{\text{work}}{\text{volume}} = \left(\frac{F_1}{A} \times \frac{x_1}{L}\right) + \left(\frac{F_2}{A} \times \frac{x_2}{L}\right) + \left(\frac{F_3}{A} \times \frac{x_3}{L}\right) + \dots$$

But the term in each bracket is just the product of the stress and strain. Thus the work done per unit volume of material is the area under the stress–strain graph up to the strain corresponding to extension $x$.

Thus when a material is stretched or compressed, work has to be done. This involves the transfer of energy into the material. Energy becomes stored in the material when it is in the stretched state. When the material is released, this stored energy is released. There is thus a danger that this release of energy with, for example, a stretched length of wire can cause the wire to whip back very rapidly.

### 2.2.5 Ductility/brittleness

If you drop a glass and it breaks, then it is possible to stick all the pieces together again and restore the glass to its original shape. The glass is said to be a *brittle* material. If a car is involved in a collision, the bodywork is less likely to shatter like the glass but more likely to dent and show permanent deformation, i.e. the material has shown plastic deformation. The term *permanent deformation* is used for changes in dimensions that are not removed when the forces applied to the material are removed. Materials which develop significant permanent deformation before they break are called *ductile* materials. The carbody work is likely to be mild steel, as such a material is a ductile material. Figure 2.11 shows the types of stress–strain graphs given by brittle and ductile materials, the ductile one showing a considerable extent of plastic behaviour. Thus if we consider the stress–strain graphs given in Figure 2.8, cast iron and glass are brittle materials, mild steel and polythene are not.

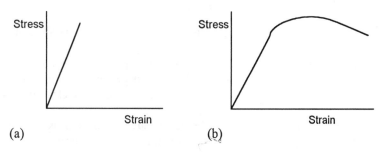

Figure 2.11   *Stress-strain graphs to fracture: (a) brittle matrials; (b) ductile materials*

Ductile materials permit manufacturing methods which involve bending them to the required shapes or using a press to squash the material into the required shape. Brittle materials cannot be formed to shape in this way.

A measure of the ductility of a material is obtained from measurements of the length of a test piece of the material before it is stretched, then stretch it until it breaks, put the pieces together and measure the final length of the test piece, as illustrated in Figure 2.12. A brittle material will show little change in length from that of the original test piece, a ductile material will, however, show a significant increase in length. The measure of the ductility is then the *percentage elongation*, i.e.:

$$\% \text{ elongation} = \frac{\text{final} - \text{initial lengths}}{\text{initial length}} \times 100$$

A reasonably ductile material, such as mild steel, will have a percentage elongation of about 20%, or more. A brittle material, such as a cast iron, will have a percentage elongation of less than 1%. Thermoplastics tend to have percentage elongations often of the order of 50 to 500%, thermosets if the order of 0.1 to 1%. Thermosets are brittle materials, thermoplastics generally not.

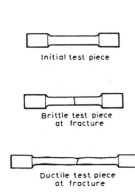

Initial test piece

Brittle test piece at fracture

Ductile test piece at fracture

Figure 2.12 *Brittle and ductile test pieces after fracture*

### Example

A material has a percentage elongation of 10%, by how much longer will a strip of the material of initial length 200 mm be when it breaks?

The percentage elongation can be expressed as:

$$\% \text{ elongation} = \frac{\text{change in length}}{\text{original length}} \times 100$$

Thus:

$$\text{change in length} = \frac{10 \times 200}{100} = 20 \text{ mm}$$

### Example

Which of the following materials is the most ductile?

80-20 brass percentage elongation 50%
70-30 brass percentage elongation 70%
60-40 brass percentage elongation 40%

The most ductile material is the one with the largest percentage elongation, i.e. the 70-30 brass.

**Example**

A sample of a carbon steel has a tensile strength of 400 MPa and a percentage elongation of 35%. A sample of an aluminium–manganese alloy has a tensile strength of 140 MPa and a percentage elongation of 10%. What does this data tell you about the mechanical behaviour of the materials?

The higher value of the tensile strength of the carbon steel indicates that the material is stronger and for the same cross-sectional area, a bar of carbon steel could withstand higher tensile forces than a corresponding bar of the aluminium alloy. The higher percentage elongation of the carbon steel indicates that the material has a greater ductility than the aluminium alloy. Indeed the value is such as to indicate that the carbon steel is very ductile.

**Example**

A material is required that can be pressed into a former in order to produce kitchen sink unit surfaces. What type of material will be required?

In order to withstand the forces involved and give the permanent deformations required, a ductile material is necessary for the task.

### 2.2.6 Toughness

The materials in many products may contain cracks or sharp corners or other changes in shape that can readily generate cracks. A tough material can be considered to be one that, though it may contain a crack, resists breaking as a result of the crack growing and running through the material. Think of trying to tear a sheet of paper or a sheet of some cloth. If there is an initial 'crack' then the material is much more easily torn. In the case of the paper, the initial 'cracks' may be perforations put there to enable the paper to be torn easily. In the case of a sheet of cloth, it may be the initial 'nick' cut in the selvedge by a dressmaker to enable it to be torn easily. A tough material is not required. However, in the case of, say, the skin of an aircraft where there may be holes, such as windows or their fastenings, which are equivalent to cracks, there is a need for cracks not to propagate. A tough material is required.

*Toughness* can be defined in terms of the work that has to be done to propagate a crack through a material, a tough material requiring more energy than a less tough one. The area under the stress–strain graph up to some strain is the energy required per unit volume of material to produce that strain (see Section 2.2.4). For a crack to propagate, a material must fail. Thus the area under the stress–strain graph up to the breaking point is a measure of the energy required to break unit volume of the material and so for a crack to propagate. A large area is given by a material with a large yield stress and high ductility (see the graph for the ductile material in Figure 2.11). Such materials can thus be considered to be tough.

An alternative way of considering toughness is the ability of a material to withstand shock loads. A measure of this ability to withstand suddenly applied forces is obtained by *impact tests*, such as the Charpy and Izod tests (see Chapter 3). In these tests, a test piece is struck a sudden blow and the energy needed to break it is measured. A brittle material will require less energy than a ductile material. The results of such tests are often used as a measure of the brittleness of materials.

### 2.2.7 Hardness

The *hardness of a material* is a measure of the resistance of a material to abrasion or indentation. A number of scales are used for hardness, depending on the method that has been used to measure it (see Chapter 3 for discussions of test methods). The tensile strength for a particular material is roughly proportional to the hardness (see Chapter 3). Thus the higher the hardness of a material, the higher is likely to be the tensile strength.

### 2.2.8 Wear resistance

*Wear* is the progressive loss of material from surfaces as a result of contact with other surfaces. It can occur as a result of sliding or rolling contact between surfaces or from the movement of fluids containing particles over surfaces. Because wear is a surface effect, surface treatments and coatings play an important role in improving wear resistance. Lubrication can be considered to be a way of keeping surfaces apart and so reducing wear.

Mild steel sliding or rolling on mild steel gives poor wear resistance. The wear resistance can, however, be improved by treatments which result in the contact surfaces being hardened. In designing an artificial hip-joint (see Section 1.2), low friction and high wear resistance are required between the surfaces of the ball and the cup parts of the joint. This can be achieved by using stainless steel for the ball and a plastic for the cup.

### 2.2.9 Fatigue

If you repeatedly flex a strip of material back and forth, it is possible to break it without the stresses every reaching the tensile or compressive stress values. This method of breaking materials is termed *fatigue* and, in some applications, materials must be chosen which have good fatigue resistance.

### 2.2.10 Density

An important physical property of a material is its density. The *density* of a material is the mass per unit volume, i.e.:

$$density = \frac{mass}{volume}$$

It has the unit of kg/m³. It is often a property that is required in conjunction with a mechanical property. Thus, for example, an aircraft undercarriage is required to be not only strong but of low mass. Thus what is required is as high a strength as possible with as low a density as possible. Thus what is looked for is a high value of strength/density. This quantity is often referred to as the *specific strength*. Steels tend to have specific strengths of about 50 to 100 MPa/Mg m⁻³ (note: 1 Mg is 10⁶ g or 1000 kg), magnesium alloys about 140 MPa/Mkg m⁻³ and titanium alloys about 250 MPa/Mkg m⁻³. Thus, for example, in some applications a lower strength magnesium alloy might be preferred to a higher strength, but higher density, steel.

## 2.3 Electrical properties

The electrical *resistivity* $\rho$ is a measure of the electrical resistance of a material, being defined by the equation

$$\rho = \frac{RA}{L}$$

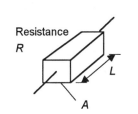

Figure 2.13 *Resistivity of a conductor*

where $R$ is the resistance of a length $L$ of a material of cross-sectional area $A$ (Figure 2.13). The unit of resistivity is the ohm metre ($\Omega$ m). An electrical insulator such as a ceramic will have a very high resistivity, typically of the order of 10¹⁰ $\Omega$ m or higher. An electrical conductor such as copper will have a very low resistivity, typically of the order of 10⁻⁸ $\Omega$ m. The term *semiconductor* is used for those materials which have resistivities roughly half way between conductors and insulators, i.e. of the order of 10² $\Omega$ m.

The electrical *conductance* $G$ of a length of material is the reciprocal of its resistance and has the unit of $\Omega^{-1}$. This unit is given a special name, the siemen (S). The electrical *conductivity* $\sigma$ is the reciprocal of the resistivity, i.e.:

$$\sigma = \frac{1}{\rho} = \frac{L}{RA} = \frac{LG}{A}$$

The unit of conductivity is thus $\Omega^{-1}\,m^{-1}$ or S/m. Since conductivity is the reciprocal of the resistivity, an electrical insulator will have a very low conductivity, of the order of 10⁻¹⁰ S/m, while an electrical conductor will have a very high conductivity, of the order of 10⁸ S/m. Semiconductors have conductivities of the order of 10⁻² S/m.

Table 2.2 shows typical values of resistivity and conductivity for insulators, semiconductors and conductors. Pure metals and many metal alloys have resistivities that increase when the temperature increases; some metal alloys do, however, show increases in resistivities when the temperature increases. For semiconductors and insulators, the resistivity increases with an increase in temperature.

Table 2.2 *Typical resistivity and conductivity values at about 20°C*

| Material | Resistivity $\Omega$ m | Conductivity S/m |
|---|---|---|
| *Insulators* | | |
| Acrylic (a polymer) | $> 10^{14}$ | $< 10^{-14}$ |
| Polyvinyl chloride (a polymer) | $10^{12}-10^{13}$ | $10^{-13}-10^{-12}$ |
| Mica | $10^{11}-10^{12}$ | $10^{-12}-10^{-11}$ |
| Glass | $10^{10}-10^{14}$ | $10^{-14}-10^{-10}$ |
| Porcelain (a ceramic) | $10^{10}-10^{12}$ | $10^{-12}-10^{-10}$ |
| Alumina (a ceramic) | $10^9-10^{12}$ | $10^{-12}-10^{-9}$ |
| *Semiconductors* | | |
| Silicon (pure) | $2.3 \times 10^3$ | $4.3 \times 10^{-4}$ |
| Germanium (pure) | $0.43$ | $2.3$ |
| Gallium arsenide | $0.05$ | $20$ |
| *Conductors* | | |
| Nichrome (alloy of nickel and chromium) | $108 \times 10^{-8}$ | $0.9 \times 10^6$ |
| Manganin (alloy of copper and manganese) | $42 \times 10^{-8}$ | $2 \times 10^6$ |
| Nickel (pure) | $7 \times 10^{-8}$ | $14 \times 10^6$ |
| Copper (pure) | $2 \times 10^{-8}$ | $50 \times 10^6$ |

**Example**

Using the value of electrical conductivity given in Table 2.2, determine the electrical conductance of a 2 m length of nichrome wire at 20°C if it has a cross-sectional area of 1 mm².

Using the equation $\sigma = L/RA$, with the conductance $G = 1/R$, then we have $\sigma = LG/A$ and so:

$$G = \frac{\sigma A}{L} = \frac{0.9 \times 10^6 \times 1 \times 10^{-6}}{2} = 0.45 \text{ S}$$

**Example**

What type of material would be needed for the heating element of an electric fire?

The heating element must be a conductor of electricity. The power dissipated by the element is $V^2/R$, thus the lower the resistance $R$ the greater the power produced by a given voltage $V$. The material must also be able to withstand high temperatures without melting or oxidising. Nichrome wire is commonly used. The wire is wound on a spiral around an insulating ceramic support.

### 2.3.1 Dielectrics

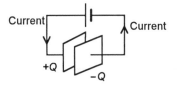

Figure 2.14 *Charging a capacitor*

When a pair of parallel conducting plates are connected to a d.c. supply Figure 2.14), charge flows on to one of the plates and off the other plate. One of the plates becomes positively charged and the other negatively charged. The amount of charge $Q$ on a plate, be it negative or positive, is proportional to the potential difference $V$ between the plates. Hence:

$$Q = CV$$

where $C$ is the constant of proportionality, called the *capacitance*. The unit of capacitance is the farad (F) when $V$ is in volts and $Q$ in coulombs. The factors determining the value of the capacitance are the plate area $A$, the separation $d$ of the plates and the medium between them:

$$C = \frac{\varepsilon A}{d}$$

where $\varepsilon$ is the factor, called the *absolute permittivity*, which relates to the medium between the plates. A more usual way of writing the equation is, however, in terms of how the permittivity of a material compares with that of a vacuum. Thus:

$$C = \varepsilon_r \varepsilon_0 \frac{A}{d}$$

where $\varepsilon = \varepsilon_r \varepsilon_0$. $\varepsilon_0$ is called the *permittivity of free space* and has a value of $8.85 \times 10^{-12}$ F/m. $\varepsilon_r$ is called the *relative permittivity*. It has no units, merely stating the factor that must be used to multiply the permittivity of free space in order to obtain the permittivity of some material. For a vaccum the relative permittivity is 1, for plastics it is between about 2 and 3, for glass between 5 and 10. The relative permittivity used to be termed the *dielectric constant* and the material between the conducting plates as the *dielectric*.

The *dielectric strength* is a measure of the highest voltage that an insulating material can withstand without electrical breakdown. It is defined as:

$$\text{dielectric strength} = \frac{\text{breakdown voltage}}{\text{insulator thickness}}$$

The units of dielectric strength are volts per metre. Table 2.3 shows some typical values.

Polyethylene has a dielectric strength of about $4 \times 10^7$ V/m. This means that a 1 mm thickness of polyethylene will require a voltage of about 40 000 V across it before it will break down.

Table 2.3 *Typical dielectric strength values*

| Material | Dielectric strength $10^7$ V/m |
|---|---|
| Polyvinyl chloride (a polymer) | 5.5 |
| Polyethylene (a polymer) | 4.0 |
| Acrylic (a polymer) | 2.0 |
| Alumina (a ceramic) | 1.2 |
| Glass | 0.2 |

**Example**

An electrical capacitor is to be made with a sheet of polythene of thickness 0.1 mm between the capacitor plates. What is the greatest voltage that can be connected between the capacitor plates if there is not to be electrical breakdown? Take the dielectric strength to be $4 \times 10^7$ V/m.

The dielectric strength is defined as the breakdown voltage divided by the insulator thickness, hence:

$$\text{breakdown voltage} = \text{dielectric strength} \times \text{thickness}$$

$$= 4 \times 10^7 \times 0.1 \times 10^{-3} = 4000 \text{ V}$$

## 2.4 Thermal properties

The S.I. unit of temperature is the kelvin (K), with a temperature change of 1 K being the same as a temperature change of 1°C. Thermal properties that are generally of interest in the selection of materials include how much a material will expand for a particular change in temperature; how much the temperature of a piece of material will change when there is a heat input into it; and how good a conductor of heat it is.

The *linear expansivity* $\alpha$ or *coefficient of linear expansion* is a measure of the amount by which a length of material expands when the temperature increases. It is defined as:

$$\alpha = \frac{\text{change in length}}{\text{original length} \times \text{change in temperature}}$$

and has the unit of $K^{-1}$.

The term *heat capacity* is used for the amount of heat (measured in joules) needed to raise the temperature of an object by 1 K. Thus if 300 J is needed to raise the temperature of a block of material by 1 K, then its heat capacity is 300 J/K. The *specific heat capacity* $c$ is the amount of heat needed per kilogram of material to raise the temperature by 1 K, hence:

$$c = \frac{\text{amount of heat}}{\text{mass} \times \text{change in temperature}}$$

It has the unit of $J\,kg^{-1}\,K^{-1}$. Because metals have smaller specific heat capacities than plastics, weight-for-weight metals require less heat to reach a particular temperature than plastics, e.g. copper has a specific heat capacity of about $340\,J\,kg^{-1}\,K^{-1}$ while polythene is about $1800\,J\,kg^{-1}\,K^{-1}$.

The *thermal conductivity* $\lambda$ of a material is a measure of the ability of a material to conduct heat. There will only be a net flow of heat energy through a length of material when there is a difference in temperature between the ends of the material. Thus the thermal conductivity is defined in terms of the quantity of heat that will flow per second divided by the temperature gradient (Figure 2.15), i.e.:

$$\lambda = \frac{\text{quantity of heat/second}}{\text{temperature gradient}}$$

and has the unit of $W\,m^{-1}\,K^{-1}$. A high thermal conductivity means a good conductor of heat. It means a small temperature gradient for a particular rate of influx of heat. Metals tend to be good conductors, for example, copper has a thermal conductivity of about $400\,W\,m^{-1}\,K^{-1}$. Materials that are bad conductors of heat have low thermal conductivities, for example, plastics have thermal conductivities of the order $0.3\,W\,m^{-1}\,K^{-1}$ or less. Very low thermal conductivities occur with foamed plastics, i.e. those containing bubbles of air. For example, foamed polymer polystyrene, known as expanded polystyrene and widely used for thermal insulation, has a thermal conductivity of about 0.02 to $0.03\,W\,m^{-1}\,K^{-1}$.

Table 2.4 gives typical values of the linear expansivity, the specific heat capacity and the thermal conductivity for metals, polymers and ceramics.

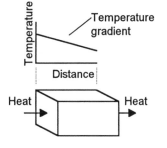

Figure 2.15 *Heat flow through a block of material*

*Table 2.4 Thermal properties*

| Material | Linear expansivity $10^{-6}\,K^{-1}$ | Specific heat capacity $J\,kg^{-1}\,K^{-1}$ | Thermal conductivity $W\,m^{-1}\,K^{-1}$ |
|---|---|---|---|
| *Metals* | | | |
| Aluminium | 24 | 920 | 230 |
| Copper | 18 | 385 | 380 |
| Mild steel | 11 | 480 | 54 |
| *Polymers* | | | |
| Polyvinyl chloride | 70–80 | 840–1200 | 0.1–0.2 |
| Polyethylene | 100–200 | 1900–2300 | 0.3–0.5 |
| Epoxy cast resin | 45–65 | 1000 | 0.1–0.2 |
| *Ceramics* | | | |
| Alumina | 8 | 750 | 38 |
| Fused silica | 0.5 | 800 | 2 |
| Glass | 8 | 800 | 1 |

**Example**

By how much will a 10 cm strip of (a) copper, (b) PVC expand when the temperature changes from 20°C to 30°C? Use the data given in Table 2.4.

The linear thermal expansivity $\alpha$ is given by:

$$\alpha = \frac{\text{change in length}}{\text{original length} \times \text{change in temperature}}$$

(a) For copper, $\alpha = 18 \times 10^{-6} \, \text{K}^{-1}$,

$$18 \times 10^{-6} = \frac{\text{expansion}}{0.10 \times 10}$$

Thus the amount of expansion is $18 \times 10^{-6}$ m or 0.018 mm.
(b) For the PVC, $\alpha = 75 \times 10^{-6} \, \text{K}^{-1}$,

$$75 \times 10^{-6} = \frac{\text{expansion}}{0.10 \times 10}$$

Thus the amount of expansion is $75 \times 10^{-6}$ m or 0.075 mm. The amount of expansion with the PVC is some four times greater than that of the copper.

**Example**

The heating element for an electric fire is wound on an electrical insulator. What thermal considerations will affect the choice of insulator material?

The insulator will need to have a low heat capacity so that little heat is used to raise the material to temperature. This means using a material with as low a density, and hence low mass, and specific heat capacity as possible. It also will need to be able to withstand the temperatures realised without deformation or melting. A ceramic is indicated.

**Example**

A designer of domestic pans requires a material for the handle which would enable a hot pan to be picked up with comfort, the handle not getting hot though the pan to which it was attached was hot. What quantity should be looked for in tables in order to find a suitable material?

What is required is a material with a very low thermal conductivity. A polymer would be suitable, provided it did not melt or deform at the temperatures realised.

## 2.5 Optical properties

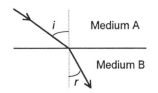

Figure 2.16 *Refraction from medium A to B*

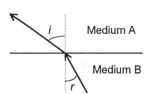

Figure 2.17 *Refraction from medium B to A*

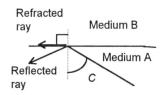

Figure 2.20 *Critical angle*

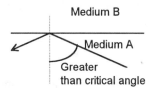

Figure 2.21 *Total internal reflection*

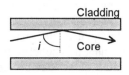

Figure 2.22 *An optical fibre*

An important optical property of a material is its refractive index. When a ray of light passes from one medium to another, for example, air into glass, reflection and refraction occur at the interface (Figure 2.16). With reflection the angle of incidence equals the angle of reflection. With refraction the ray of light bends from its straight-line path in passing across the interface. The *refractive index* in going from medium A to medium B, $_An_B$, is given by:

$$\text{refractive index } _An_B = \frac{\sin i}{\sin r}$$

where $i$ is the angle of incidence, i.e. the angle between the incident ray in medium A and the normal, and $r$ the angle of refraction, i.e. the angle between the refracted ray in medium B and the normal. This is known as *Snell's law*.

In Figure 2.16 the ray of light is shown as starting in medium A and moving into medium B, bending towards the normal. This occurs because the velocity of light in medium A is greater than that in medium B. Suppose we reverse the path and have the ray of light passing from medium B into medium A (Figure 2.17). The same path is followed, but in the reverse direction, the ray now bending away from the normal. This is because the light is passing from a medium where the speed of light is lower to one where it is higher. Thus we have, when using the same notation for the angles, a refractive index in this case of:

$$_Bn_A = \frac{\sin r}{\sin i} = \frac{1}{_An_B}$$

When a ray of light travels from a material into one in which it has a lower speed, it bends towards the normal. When a ray of light travels from a material into one in which it has a greater speed it bends away from the normal. When we have this condition of the ray bending away from the normal then we can have a particular incident angle which results in the refracted ray of light bending through 90° and thus not being transmitted across the interface (Figure 2.20). The angle of incidence in such a case is termed the *critical angle C*. We then have:

$$_Bn_A = \frac{\sin C}{\sin 90°} = \sin C$$

For angles of incidence greater than the critical angle, the ray of light is totally reflected at the interface (Figure 2.21), there being no refracted ray.

As an illustration of the significance of the critical angle in the choice of optical materials, consider the material used for *fibre optics*. The basic optical fibre consists of a central core of material in which the velocity of light is higher than in the surrounding cladding. Light for which the angle of incidence is greater than the critical angle is transmitted along such a fibre by total internal reflection, none of such light being lost from the fibre by being refracted through the cladding (Figure 2.22).

The refractive index used above is that for light travelling from one material to another and is referred to as the *relative refractive index*. For example, we thus have $_An_B$ for light travelling from medium A to medium B. The refractive index is in fact the ratio of the velocities of light in the two media:

$$_An_B = \frac{\text{velocity of light in A}}{\text{velocity of light in B}}$$

It is convenient to define an *absolute refractive index* of a medium as being that given when light travels from a vacuum into that medium, i.e.:

$$n_A = \frac{\text{velocity of light in a vacuum}}{\text{velocity of light in A}}$$

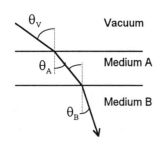

Figure 2.23  *Refraction from vacuum to A to B*

To arrive at the relationship between the absolute refrative indices of two media A and B and the relative refractive index in going from A to B, consider the situation shown in Figure 2.23 when light travels from a vacuum into medium A and then into B. At the interface between the vacuum and medium A we can write:

$$n_A = \frac{c}{c_A}$$

where $c$ is the velocity of light in a vacuum and $c_A$ that in medium A. For the interface between medium A and B we can write:

$$_An_B = \frac{c_A}{c_B}$$

This can be rewritten as:

$$_An_B = \frac{c_A}{c} \times \frac{c}{c_B} = \frac{n_B}{n_A}$$

Thus knowing the absolute refractive indices for two media enables us to calculate the relative refractive index for the two media. But we have $_An_B = \sin\theta_A/\sin\theta_B$, thus we can write Snell's law as

$$n_A \sin\theta_A = n_B \sin\theta_B$$

Light when incident on a material can be reflected, absorbed and transmitted. The transparency of a material, such as a plastic, depends on its light-absorbing and light-scattering properties. The term *total transmission factor* is used for the ratio of the total transmitted light intensity and the incident light intensity, assuming it is concentrated in a parallel beam perpendicular to the surface of the sample. For comparison purposes the values are usually quoted for a thickness of 1 mm. The *reflection factor* is the ratio of the light intensity reflected at an angle equal to the angle of incidence and the intensity of the incident beam, assuming it is concentrated in a parallel beam. The *clarity* with which detail in an object can be seen when viewed through a sample of the material depends on the amount of light scattered in the material. It is perfect only when no light is scattered.

Polyethylene typically has a refractive index of about 1.52 and a direct transmission factor, for low density polyethylene, of about 40–45%. Polyvinyl chloride has a refractive index of about 1.54 and a direct transmission factor of about 90%. This high transparency means it is frequently used as a glass substitute, it having the advantage of not breaking so readily.

**Example**

Determine the critical angle for the glass/air interface if the glass has a refractive index of 1.5.

The refractive index of the glass is for light going from air to glass. Thus:

$$1.5 = \frac{\sin 90°}{\sin C}$$

Hence the critical angle $C$ is 41.8°.

**Example**

An optical fibre consists of a glass core clad with another material. The core has an absolute refractive index of 1.40 and the cladding an absolute refractive index of 1.42. What is the critical angle for light incident on the glass/cladding interface?

The refractive index for light passing from glass to the cladding is:

$$_g n_c = \frac{n_c}{n_g} = \frac{1.40}{1.42} = 0.99$$

Thus the critical angle $C$ is given by:

$$0.99 = \frac{\sin C}{\sin 90°}$$

Hence the critical angle is 81.9°.

## 2.6 Chemical properties

Attack of materials by the environment in which they are situated can be a major problem. The rusting of iron in air is an obvious example of such an attack. Tables are available giving the comparative resistance to attack of materials in various environments, e.g. in aerated water, in salt water, to strong acids, strong alkalis, organic solvents and ultraviolet radiation.

Selection of an appropriate material for a specific environment can do much to reduce such attack. For example, in a salt water environment, carbon steels are rated as having very poor resistance to attack, aluminium alloys good resistance and stainless steels excellent resistance. Aluminium readily develops a durable protective surface film which then increases the resistance of the material to attack. In the case of steel, the addition of chromium to it can considerable improve its resistance to attack by modifying the surface protective film produced under oxidising conditions. The material is referred to as stainless steel. Attack can often be prevented or reduced by modification of the environment to which a material is exposed. This might be by painting it or putting it in a protective package. When the environment adjacent to the material is liquid, it is often possible to add certain compounds to the liquid so that attack is inhibited, such additives being termed *inhibitors*. In the case of water in steel radiators or boilers, compounds which provide chromate or phosphate ions may be used as inhibitors. They help to maintain protective surface films on the steel.

While some polymers are highly resistant to chemical attack, others are liable to stain, craze, soften, swell or dissolve completely. For example, nylon shows little degradation with weak acids but is attacked by strong acids; it is resistant to alkalis and organic solvents. Polymers have generally high resistance to attack in water and thus are widely used for containers and pipes. Polymers are generally affected by exposure to sunlight. Ultra-violet light, present in sunlight, can cause a breakdown of the bonds in the polymer molecular chains and result in surface cracking. For this reason, plastics often have a ultraviolet inhibitor mixed with the polymer when the material is produced.

## 2.7 Magnetic properties

In the vicinity of permanent magnets and current-carrying conductors, a magnetic field is said to exist. The magnetic field pattern can be plotted using a compass needle or demonstrated by scattering iron filings in the vicinity. The term *magnetic line of force* is used for a line traced out by such plotting or the iron filings. A useful way of considering magnetic fields is in terms of *magnetic flux*, this being something that is considered to flow along these lines of force like water through pipes. The term *magnetic flux density B* is used for the amount of flux passing through unit area (Figure 2.24). Thus if flux $\Phi$ passes through an area $A$, then:

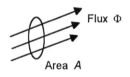

Figure 2.24 *Flux density*

$$B = \frac{\Phi}{A}$$

Magnetic flux is produced within magnetic materials when electrical currents pass through coils of wire wrapped round cores of such materials. The flux density produced for a given number of turns on a coil and a given current depends on the magnetic material used for the core. The term *relative permeability* $\mu_r$ is used for the unitless factor by which the flux density in a material $B$ compares with that which would have been produced with a vacuum as the core ($B_0$):

$$\mu_r = \frac{B}{B_0}$$

The relative permeability for air is about 1, since air is virtually the same as a vacuum. For iron the relative permeability is typically about 2000 to 10 000, though special steels can have values of the order of 60 000 to

90 000. The relative permeability for a particular material is not constant, depending on the size of magnetising field used.

When an initially unmagnetised material is placed in an increasing magnetic field, the flux density within the material increases in the way shown in Figure 2.25. As will be apparent from an inspection of Figure 2.25, the ratio $B/B_0$ is not a constant. If the magnetic field is then reduced back to zero, the material may not simply just retrace its path back down the same graph and may retain some magnetism when the applied magnetic field is zero. The retained flux density is termed the *remanent flux density* or *remanence*. To demagnetise the material, i.e. bring $B$ to zero, a reverse field called the *coercive flux density* or *coercivity* must be applied. Figure 2.26 shows how the flux density $B$ within the material might vary when an alternating magnetic field is applied, i.e. the magnetic field increased to some value, then decreased to zero, then increased to some value in the opposite direction, then decreased to zero, etc. The variation of the flux density $B$ within a material with the applied flux density $B_0$ is called a *hysteresis loop.*

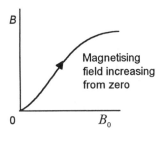

Figure 2.25 *Initial magnetisation graph*

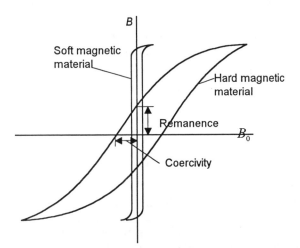

Figure 2.26 *Hysteresis loops for a soft and a hard magnetic material*

In Figure 2.26 the hysteresis loops are shown for two materials, termed hard and soft magnetic materials. The remanence and the coercivity are shown for the hard magnetic material. Compared with a soft magnetic material, a hard magnetic material has high remanence so that a high degree of magnetism is retained in the absence of a magnetic field, a high coercivity so that it is difficult to demagnetise and a large area enclosed by the hysteresis loop. The area of the loop is related to the energy dissipated in the material during each cycle of magnetisation. A soft material is very easily demagnetised, having low coercivity and the hysteresis loop only enclosing a small area. Hard magnetic materials are used for such applications as permanent magnets while soft magnetic materials are used for transformers where the magnetic material should be easily demagnetised and liittle energy dissipated in magnetising it.

**2.8 Properties in general**
Table 2.5 summarises the properties of metals, polymers and ceramics. The table does not include composites, these are discussed in Chapter 4.

Table 2.5 *The range of properties*

| Property | Metals | Polymers | Ceramics |
| --- | --- | --- | --- |
| Density Mg/m³ | Medium–high 2–16 | Low 1–2 | Generally medium 2–4 |
| Melting point °C | Medium–high 200–3500 | Low 70–200 | High 2000–4000 |
| Thermal conductivity | High | Low | Medium–low |
| Thermal expansion | Medium | High | Low |
| Specific heat capacity | Low | Medium | High |
| Electrical conductivity | High | Very low | Very low |
| Optical properties | Opaque | Some transparent, some opaque | Some transparent, some opaque |
| Tensile strength MPa | Medium–high 100–2500 | Generally low* 30–80 | Generally low 10–400 |
| Compressive strength MPa | Medium–high, as tensile | Generally low*, as tensile | High 1000–5000 |
| Tensile modulus GPa | Medium–high 40–400 | Low* 0.1–4 | High 150–450 |
| Toughness | Good | Some good, some poor* | Poor |
| Hardness | Medium | Low | High |
| Wear resistance | Medium | Low–moderate | High |
| Resistance to corrosion | Medium–poor | Good–medium | Good |

Note: 1 Mg/m³ = 1000 kg/m³. * Polymers are widely used with fillers, such as fibres and particles, and these can markedly change their properties, in particular making them stiffer, stronger and tougher.

**Example**

Which type of material, metal, polymer or ceramic, would be the most likely to give materials with each of the following properties:

(a) High density.
(b) High melting point.
(c) High electrical conductivity.
(d) Low specific heat capacity.
(e) Low tensile modulus of elasticity.
(f) High thermal conductivity.
(g) High wear resistance.

(a) Metals contain the materials with the highest densities.
(b) The highest melting points are given by the ceramics.
(c) The highest electrical conductivities are given by the metals.
(d) The lowest specific heat capacities are given by the metals.
(e) Polymers give the materials with the lowest tensile modulus of elasticity.
(f) The highest thermal conductivities are given by metals.
(g) Ceramics give the materials with the highest wear resistance.

### 2.8.1 Costs

Costs can be considered in relation to the basic costs of the raw materials, the costs of manufacturing products, and the life and maintenance costs of the finished product.

In comparing the basic costs of materials, the comparison is often on the basis of the cost per unit weight or cost per unit volume. Thus, for example, if the cost of 10 kg of a metal is, say, £1, then the cost per kg is £0.1. If the metal has a density of 8000 kg/m$^3$ then 10 kg will have a volume of $10/8000 = 0.001\ 25$ m$^3$ and so the cost per cubic metre of the material is $1/0.001\ 25 = £800$. Thus we can write

$$\text{cost per m}^3 = (\text{cost/kg}) \times \text{density}$$

However, often a more important comparison is on the basis of the cost per unit strength or cost per unit stiffness for the same volume of material. This enables the cost of, say, a beam, to be considered in terms of what it will cost to have a beam of a certain strength or stiffness. Hence if, for comparison purposes, we consider a beam of volume 1 m$^3$ then, if the tensile strength of the material is 500 MPa and the cost per cubic metre £800, the cost per MPa of strength will be $800/500 = £1.6$. Thus we can write, for the same volume of material,:

$$\text{cost per unit strength} = \frac{(\text{cost/m}^3)}{\text{strength}}$$

The example that follows illustrates such a calculation. Similarly we can write a cost per unit stiffness as:

$$\text{cost per unit stiffness} = \frac{(\text{cost/m}^3)}{\text{modulus}}$$

The costs of manufacturing will depend on the processes used. Some processes require a large capital outlay and then can be used to produce large numbers of the product at a relatively low cost per item. Other processes may have little in the way of setting-up costs but a large cost per unit product.

The cost of maintaining a material during its life can often be a significant factor in the selection of materials. A feature common to many metals is the need for a surface coating to protect them for corrosion by the atmosphere. The rusting of steels is an obvious example of this and dictates the need for such activities as the continuous repainting of the Forth Railway Bridge.

### Example

On the basis of the following data, compare the costs per unit strength of the two materials for the same volume of material.

Low carbon steel:
   Cost per kg £0.1, density 7800 kg/m³, strength 1000 MPa
Aluminium–manganese alloy:
   Cost per kg £0.22, density 2700 kg/m³, strength 200 MPa

For the steel, the volume of 1 kg is 1/7800 = 0.000 128 m³ and so the cost per m³ is 0.1/0.000128 = £780. The cost per MPa of strength is thus 780/1000 = £0.78. For the aluminium alloy, the volume of 1 kg is 1/2700 = 0.000 37 m³ and so the cost per m³ is 0.22/0.000 37 = £590. Thus although the cost per kg for the aluminium alloy is greater than that of the steel, because of the lower density, the cost per cubic metre is less. For the aluminium alloy, the cost per MPa of strength is thus 590/200 = £2.95. Hence on a comparison of the strengths of equal volumes, it is cheaper to use the steel.

## Problems

*Questions 1 to 22 have four answer options: A, B, C and D. Choose the correct answer from the answer options.*

1 Two wires of the same material are stretched by the same size force. Wire X has twice the cross-sectional area of the other wire, Y. Which one of the following statements *cannot* be correct?

   A Wire X stretches less than wire Y
   B Wire X is subject to less stress than wire Y
   C Wire X shows less strain than wire Y
   D Wire X has a smaller value of Young's modulus than wire Y

Questions 2 to 5 relate to Figure 2.27. This shows the stress–strain graph for four materials.

2 Which of the materials is the most ductile?
3 Which of the materials is the most brittle?
4 Which of the materials is the strongest?
5 Which of the materials is the stiffest?

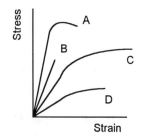

Figure 2.27 *Stress–strain graph for four materials*

Questions 6 to 10 relate to the following information.

The following are a number of different types of materials:

A   A metal.
B   A polymer.
C   An amorphous ceramic.
D   A glass.

6   Which one of the above materials might be expected to be the least stiff?
7   Which one of the above materials might be expected to have the highest electrical conductivity?
8   Which one of the above materials is likely to have the highest density?
9   Which one of the above materials is likely to be the hardest?
10   Which one of the above materials might be expected to have the highest thermal conductivity?

11   A number of parallel plate capacitors are to be made by inserting sheets of the following materials, between and in contact with two metal plates.

| Material | Dielectric strength $10^6$ V/m |
|---|---|
| Mica | 60 |
| Polyethylene | 20 |
| Dry paper | 16 |
| Electrical ceramic | 12 |

Which one of the following capacitors will have the highest breakdown voltage?

A   Made with a mica sheet of thickness 0.5 mm.
B   Made with a polyethylene sheet of thickness 0.5 mm.
C   Made with a sheet of dry paper of thickness 2.0 mm.
D   Made with a sheet of electrical ceramic of thickness 3.0 mm.

12   Decide whether each of the following statements is TRUE (T) or FALSE (F).

The resistance of a wire of a particular material is greater:
(i)   The longer the wire.
(ii)   The greater the cross-sectional area of the wire.

A   (i) T   (ii) T
B   (i) T   (ii) F
C   (i) F   (ii) T
D   (i) F   (ii) F

**13** Decide whether each of the following statements is TRUE (T) or FALSE (F).

X and Y are two identical volume blocks of different materials and different masses. When there is the same heat input to both blocks, the greatest change in temperature will occur for the block having the material with the:
(i) Highest value of specific heat capacity × density.
(ii) Highest value of 1/(specific heat capacity × density).

A (i) T (ii) T
B (i) T (ii) F
C (i) F (ii) T
D (i) F (ii) F

**14** A sheet of a polymer of thickness 1 mm transmits 60% of the light intensity incident on it. A sheet of the same material 3 mm thick will thus transmit:

A 60% of the incident light
B 30% of the incident light
C 20% of the incident light
D 15% of the incident light

**15** Decide whether each of the following statements is TRUE (T) or FALSE (F).

A ray of light enters a parallel sheet of glass. The refractive index for light passing from air into the glass is 1.5.
(i) On entering the glass, the ray is refracted towards the normal.
(ii) On leaving the opposite side of the sheet of glass, the ray is refracted away from the normal.

A (i) T (ii) T
B (i) T (ii) F
C (i) F (ii) T
D (i) F (ii) F

**16** Decide whether each of the following statements is TRUE (T) or FALSE (F).

Figure 2.28 shows a stress–strain graph for a material. This shows that when a load is applied to a strip of the material:
(i) It will extend as the load is increased but eventually increasing the load will produce no increase in extension.
(ii) The stiffness of the material will increase as the load increases.

A (i) T (ii) T
B (i) T (ii) F
C (i) F (ii) T
D (i) F (ii) F

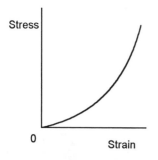

Figure 2.28 *Stress–strain graph*

**17** Decide whether each of the following statements is TRUE (T) or FALSE (F).

The material used for clothing needs to have:
(i)  A high toughness.
(ii) A high modulus of elasticity.

A  (i) T  (ii) T
B  (i) T  (ii) F
C  (i) F  (ii) T
D  (i) F  (ii) F

**18** Decide whether each of the following statements is TRUE (T) or FALSE (F).

The material used for an electrical conductor needs to have:
(i)  A high ductility.
(ii) A high electrical resistivity.

A  (i) T  (ii) T
B  (i) T  (ii) F
C  (i) F  (ii) T
D  (i) F  (ii) F

**19** Decide whether each of the following statements is TRUE (T) or FALSE (F).

The material used for the spring used with bathroom scales needs to have:
(i)  A high ductility.
(ii) A high elastic limit.

A  (i) T  (ii) T
B  (i) T  (ii) F
C  (i) F  (ii) T
D  (i) F  (ii) F

**20** Decide whether each of the following statements is TRUE (T) or FALSE (F).

The material used for the walls of an aerosol can (a pressure vessel) needs to have:
(i)  A high yield stress.
(ii) A high toughness.

A  (i) T  (ii) T
B  (i) T  (ii) F
C  (i) F  (ii) T
D  (i) F  (ii) F

**21** Decide whether each of the following statements is TRUE (T) or FALSE (F).

A hard magnetic material is one that has a hysteresis loop:
(i) Enclosing a large area.
(ii) With a high coercivity.

A (i) T (ii) T
B (i) T (ii) F
C (i) F (ii) T
D (i) F (ii) F

**22** Decide whether each of the following statements is TRUE (T) or FALSE (F).

A magnetic material is required for use as a transformer core. This material is required to be easily demagnetised and require little energy to become magnetised. Such a material has:
(i) A high coercivity.
(ii) An hysteresis loop enclosing a large area.

A (i) T (ii) T
B (i) T (ii) F
C (i) F (ii) T
D (i) F (ii) F

**23** What types of properties would be required for the following products?
(a) A domestic kitchen sink.
(b) A shelf on a bookcase.
(c) A cup.
(d) An electrical cable.
(e) A coin.
(f) A car axle.
(g) The casing of a telephone.

**24** For each of the products listed in problem 23, identify a material that is commonly used and explain why its properties justify its choice for that purpose.

**25** Which properties of a material would you need to consider if you required materials which were:
(a) Stiff.
(b) Capable of being bent into a fixed shape.
(c) Capable of not fracturing when small cracks are present.
(d) Not easily breaking.
(e) Acting as an electrical insulator.
(f) A good conductor of heat.
(g) Capable of being used as the lining for a tank storing acid.

**26** A colleague informs you that a material has a high tensile strength with a low percentage elongation. Explain how you would expect the material to behave.

**27** A colleague informs you that a material has a high tensile modulus of elasticity and good fracture toughness. Explain how you would expect the material to behave.

**28** What is the tensile stress acting on a strip of material of cross-sectional area 50 mm$^2$ when subject to tensile forces of 1000 N?

**29** Tensile forces act on a rod of length 300 mm and cause it to extend by 2 mm. What is the strain?

**30** An aluminium alloy has a tensile strength of 200 MPa. What force is needed to break a bar of this material with a cross-sectional area of 250 mm$^2$?

**31** A test piece of a material is measured as having a length of 100 mm before any forces are applied to it. After being subject to tensile forces, it breaks and the broken pieces are found to have a combined length of 112 mm. What is the percentage elongation?

**32** A material has a yield stress of 250 MPa. What tensile forces will be needed to cause yielding if the material has a cross-sectional area of 200 mm$^2$?

**33** A sample of high tensile brass is quoted as having a tensile strength of 480 MPa and a percentage elongation of 20%. An aluminium-bronze is quoted as having a tensile strength of 600 MPa and a percentage elongation of 25%. Explain the significance of this data in relation to the mechanical behaviour of the materials.

**34** A grey cast iron is quoted as having a tensile strength of 150 MPa, a compressive strength of 600 MPa and a percentage elongation of 0.6%. Explain the significance of the data in relation to the mechanical behaviour of the material.

**35** A colleague states that he/she needs a material with a high electrical conductivity. However, only electrical resistivity tables for materials are available. What type of resistivity values would you suggest he/she looks for?

**36** Aluminium has a resistivity of $2.5 \times 10^{-8}$ $\Omega$ m. What will be the resistance of an aluminium wire with a length of 1 m and a cross-sectional area of 2 mm$^2$?

**37** You read in a textbook that: 'Designing with ceramics presents problems that do not occur with metals because of the almost complete absence of ductility with ceramics.' Explain the significance of the comment in relation to the exposure of ceramics to forces.

38 Compare the specific strengths, and costs per unit strength for equal volumes, for the materials giving the following data:

Low carbon steel:
Cost per kg £0.1, density 7800 kg/m$^3$, strength 1000 MPa.
Polypropylene:
Cost per kg £0.2, density 900 kg/m$^3$, strength 30 MPa.

# 3 Materials testing

## 3.1 Determination of properties

This chapter is about the determination of some of the basic properties of materials. The approach adopted is to describe experiments with simple apparatus that can be carried out in the typical school/college laboratory and also examples of the tests that are used in industry. In order that the results of industrial tests are readily interpreted by industrial users of materials, the test will generally be carried out and the data presented in a form laid down by the British Standards Institution or other national or international standards authorities.

The following list illustrates the types of specifications of standard tests issued by the British Standards Institution.

| | |
|---|---|
| BSEN 10002 formerly BS18 | Methods of tensile testing of metals |
| BS 131 | Methods of notched bar tests |
| BSEN 10045 | Charpy test |
| BS 240 | Method for Brinell hardness test |
| BS 427 | Method for Vickers hardness test |
| BS 891 | Method for Rockwell hardness test |
| BS 1639 | Method of bend testing of metals |
| BS EN 2155: Part 5 | Test methods for the determination of the light transmission of aircraft glazing |
| BS 2782 | Methods of testing plastics |
| BS 5714 | Resistivity measurements for metals |
| BS 7030 | Determination of the coefficient of expansion of glass |

Standards that are specified as BSEN are European standards which have been adopted as British standards.

## 3.2 Errors

All measurement systems contain sources of error that limit the accuracy of data obtained from measurements. The following are some of the common sources of error:

1   *Construction errors*
    These result from causes such as tolerances on the dimensions of components and the values of electrical components used, and are inherent in the manufacture of an instrument. In addition, there can be errors due to the accuracy with which the manufacturer of an instrument has calibrated it.

2   *Approximation errors*
In the design of many instrument it is often the case that a linear relationship between two quantities is assumed, e.g. with a spring balance, a linear relationship between force and extension. This may be an approximation or may be restricted to a narrow range of values. Thus an instrument may have errors due to a component not having a perfectly linear relationship.

3   *Operating errors*
These can occur for a variety of reasons. They may occur in reading the position of a pointer on a scale. If the scale and the pointer are not in the same plane, then the reading obtained depends on the angle at which the pointer is viewed against the scale (Figure 3.1). These are called *parallax errors*. To reduce the chance of such errors, some instruments incorporate a mirror alongside the scale so that the pointer is seen both against a scale reading and reflected in the mirror. By positioning the eye so that the pointer and its image are in line guarantees that the pointer is being viewed at the right angle. Digital instruments, where the reading is displayed as a series of numbers, avoid this problem of parallax. Errors may also occur due to the limited resolution of an instrument and taking a reading from a scale. Operating errors can also arise when an instrument has to be brought into contact with an object being measured, e.g. a micrometer, as a result of slightly different contact forces. There are also *human errors*, e.g. such as reading the wrong scale.

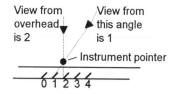

Figure 3.1 *Parallax errors*

4   *Environmental errors*
Errors can arise as a result of environmental effects, e.g. a change in temperature affecting the value of a resistance.

5   *Insertion errors*
In some measurements the insertion of the instrument into the position to measure a quantity can affect its value. For example, inserting an ammeter into a circuit to measure the current can affect the value of the current due to the ammeter's own resistance. Similarly, putting a cold thermometer into a hot liquid can cool the liquid and so change the temperature being measured.

One particular form of operating error associated with the scale markings on an instrument is the *reading error*. When the pointer of an instrument falls between two scale markings (Figure 3.2) there is some degree of uncertainty as to what the reading should be quoted as. Thus the reading should not be quoted as a precise number but some indication given of the possible extent to which the reading could be in error. The worst the reading error could be is that the value indicated by a pointer is somewhere between two successive markings on the scale. In such circumstances, the reading error can be stated as a value ± half the scale interval. However, it is often the case that we can be more certain about the reading and indicate a smaller error. With digital displays, there is no uncertainty regarding the

Figure 3.2 *Reading error*

value displayed but there is still an error associated with the reading. This is because the reading of the instrument goes up in jumps, a whole digit at a time. We cannot tell where between two successive digits the actual value really is. Thus the degree of uncertainty is ± the smallest digit.

### 3.2.1 Random and systematic errors

All errors, whatever their source, can be described as being either random or systematic errors. *Random errors* are ones that can vary in a random manner between successive readings of the same quantity e.g. those due to parallax errors. *Systematic errors* are errors that do not vary from reading to another, e.g. those due to some defect in the instrument such as a wrongly set zero so that it always gives a high or low reading. Random errors can be minimised by the use of statistical analysis, as illustrated below. Systematic errors can be detected and eliminated by the use of a different instrument or measurement technique, on the assumption that the systematic errors will not be the same in both cases.

### 3.2.2 Distribution of values with random errors

Random errors mean that sometimes the error will give a reading that is sometimes or too low. The error can be reduced by taking repeated readings and calculating the *average* (or *mean*) value. The average $\bar{x}$ of a set of readings is given by:

$$\bar{x} = \frac{x_1 + x_2 + \ldots x_n}{n}$$

where $x_1$ is the first reading, $x_2$ the second reading, .... $x_n$ the $n$th reading. We can be more certain about an average value than just the value obtained by one measurement. The more readings we take, the more likely it will be that we can cancel out the random variations that occur between readings. The *true value* might thus be regarded as the value given by the average of a very large number of readings.

Consider the following set of 12 values obtained for some quantity:

10, 10, 9, 10, 8, 12, 11, 10, 10, 11, 10, 9

The average is 120/12 = 10. We can show how the values are spread around the average value by drawing a *frequency distribution*. To do this we need to determine the frequency with which values occur. In this case we have:

8 once, 9 twice, 10 five times, 11 twice, 12 once

Figure 3.3 shows the resulting frequency distribution.

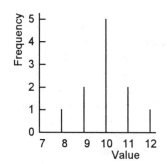

Figure 3.3 *Frequency distribution*

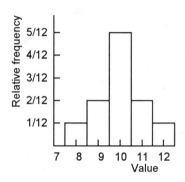

Figure 3.4 *Histogram*

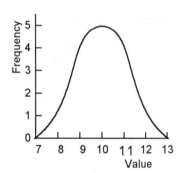

Figure 3.5 *Continuous frequency distribution*

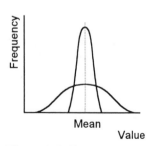

Figure 3.6 *Frequency distributions*

Quite often it is not appropriate to use discrete values but to determine the frequency with which values occur in certain ranges of values. For example, the values given above might be determined with more precision as:

10.1, 9.9, 8.8, 10.2, 7.9, 12.1, 11.3, 9.7, 10.0, 11.0, 10.4, 8.6

The average is 120.0/12 = 10.0. If we group the data values $x$ into groups where the bottom value of a group is $x_1$ and the upper value $x_2$ and the group is specified by $x_1 < x \le x_2$, then we have for each interval:

7.5 to 8.5, once
8.5 to 9.5, twice
9.5 to 10.5, five times
10.5 to 11.5, twice
11.5 to 12.5, once

The *relative frequency* with which values occur in each group is the frequency for that group divided by the total number of values. Thus we have:

7.5 to 8.5, relative frequency 1/12
8.5 to 9.5, relative frequency 2/12
9.5 to 10.5, relative frequency 5/12
10.5 to 11.5, relative frequency 2/12
11.5 to 12.5, relative frequency 1/12

The histogram shows the relative frequency values as rectangles based on each group of values, the area of the rectangle being equal to the relative frequency. With equal size groups of values the width of each rectangle is the same. Thus for the data given above, the histogram is as shown in Figure 3.4.

If we have a large number of results then we can draw a frequency distribution or a histogram with quite small intervals and it is then feasible to think of a smooth curve drawn through the outline of the histogram. The result might then be of the form shown in Figure 3.5.

Suppose the same physical quantity is measured in two different experiments which involve different apparatus. We might obtain the frequency distributions shown in Figure 3.6. Both experiments give the same average value, but with one experiment, the results are scattered less about the mean value than the other. The experiment giving the least spread is said to have greater *precision* than the other experiment. The higher precision experiment gives results with small random errors. The *deviation* of any one reading from the mean is the difference between its value and the mean value, thus the less precise set of readings has greater deviations than the more precise set. The spread of the readings is specified by a quantity termed the *standard deviation*, where:

$$\text{standard deviation} = \sqrt{\frac{\left(d_1^2 + d_2^2 + \dots d_n^2\right)}{n-1}}$$

where $d_1$ is the deviation of the first result from its average, $d_2$ the deviation of the second reading, ... $d_n$ the deviation of the $n$th reading from the average. Note that sometimes the above equation is written with just $n$ on the bottom line. This is because with large numbers of readings $(n-1)$ is virtually the same as $n$. Generally, for the spread of readings encountered with experimental measurements, we can reasonably expect with large numbers of readings that about 68.3% of the readings will lie within plus or minus one standard deviation of the mean value, 95.5% within plus or minus two standard deviations, 99.7% within plus or minus three standard deviations and 99.99% within plus or minus four standard deviations.

The standard deviation lets us know how far from the mean value we can expect any one reading to be. The chance of a measurement falling within $0.6745\sigma$ of the true value is 50%. The $0.6745\sigma$ is called the *probable error*. If we consider a measurement, then the chance that it will lie within plus or minus one standard deviation of the mean value is 68.3%. Another way of expressing this is that the limits $\pm 1\sigma$ for a measurement give the 68.3% *confidence limits of an interval* within which a measurement will be expected to fall. Similarly the limits $\pm 2\sigma$ give a 95.5% confidence interval. A *X% confidence interval* for some quantity is defined as the interval within which the chance of a measurement falling is $X/100$.

In an experiment, we might take a relatively small number of readings. The average value obtained from such a set of a small number of readings might be in error, i.e. differ from the true value that would be obtained from a very large number of readings. The standard deviation tells us how far from the true value we might expect a particular reading to occur. However, if we take a number of readings, what we want to know is how far the average of the set of readings is likely to be from the true value. What we want is the standard deviation of the mean of our set of readings from the true value, i.e. the mean value that would occur with a large number of readings. This is given by what is termed the *standard error of the mean*,

$$\text{standard error} = \frac{\text{standard deviation}}{\sqrt{n}}$$

Thus the more readings we take, the closer we might expect our mean to be to the true value, i.e. the smaller the standard error is likely to be. We can talk of a confidence interval for the standard error. As with the standard deviation, the limits $\pm 1$(standard error) gives the 68.3% *confidence limits of an interval* within which the mean will be expected to fall. Similarly the limits $\pm 2$ (standard error) give a 95.5% confidence interval.

**Example**

Measurements of the electrical resistance of a resistor by the use of the ammeter–voltmeter method gave the following results, in ohms:

53, 48, 45, 49, 46, 48, 51, 57, 55, 55, 47, 49

Determine the average value for the resistance, and the standard error of the average.

Table 3.1 *Example*

| Resistance $\Omega$ | Deviation $\Omega$ | (Deviation)$^2$ $\Omega^2$ |
|---|---|---|
| 53 | +2.75 | 7.5625 |
| 48 | −2.25 | 5.0625 |
| 45 | −5.25 | 27.5625 |
| 49 | −1.25 | 1.5625 |
| 46 | −4.25 | 18.0625 |
| 48 | −2.25 | 5.0625 |
| 51 | +0.75 | 0.5625 |
| 57 | +6.75 | 45.5625 |
| 55 | +4.75 | 22.5625 |
| 55 | +4.75 | 22.5625 |
| 47 | −2.25 | 5.0625 |
| 49 | −1.25 | 1.5625 |

Table 3.1 outlines the steps involved. The average is the sum of the values in the first column divided by the number of values involved, i.e. $603/12 = 50.25\ \Omega$. The deviations of each of the readings from this average value are: $53 - 50.25 = +2.75$, $48 - 50.25 = -2.25$, etc. and are given in the second column. The squares of these deviations are given in the third column. The sum of these squares is 168.25. Hence the standard deviation is $\sqrt{(168.25/11)} = 3.91$. The standard error is thus $3.91/\sqrt{12} = 1.13\ \Omega$. Thus we can write our estimate of the resistance as $50.25 \pm 1.13\ \Omega$.

**Example**

The mean of 12 measurements has been found to be 12.9 and the standard deviation 1.5. Determine the 95% confidence interval for this mean.

A 95% confidence interval about the mean value is given by about ±2 (standard error). The standard error is $\sigma/\sqrt{n}$, where $\sigma$ is the standard deviation and $n$ the number of readings taken. Thus

$$\text{confidence interval} = 12.9 \pm \frac{2 \times 1.5}{\sqrt{12}} = 12.9 \pm 0.9$$

We can be 95% certain that the mean of 12.9 will fall within this interval of the true value.

### 3.2.3 Handling errors

Consider the calculation of the quantity $Z$ from two measured quantities $A$ and $B$ where $Z = A + B$. If the measured quantity $A$ has an error $\pm \Delta A$ and the quantity $B$ an error $\pm \Delta B$, then the worst possible error in $Z$ is if the two errors are both positive or both negative, then we have:

$$Z + \Delta Z = A + \Delta A + B + \Delta B$$

$$Z - \Delta Z = A - \Delta A + B - \Delta B$$

Thus the worst possible error in $Z$ is:

$$\Delta Z = \Delta A + \Delta B$$

Thus when we add two measured quantities the worst possible error in the calculated quantity is the sum of the errors in the measured quantities.

If we have the calculated quantity $Z$ as the difference between two measured quantities, i.e. $Z = A - B$, then, in a similar way, we can show that the worst possible error is given by:

$$\Delta Z = \Delta A + \Delta B$$

Thus when we subtract two measured quantities the worst possible error in the calculated quantity is the sum of the errors in the measured quantities.

If we have the calculated quantity $Z$ as the product of two measured quantities $A$ and $B$, i.e. $Z = AB$, then we can calculate the worst error in $Z$ as being when the errors in $A$ and $B$ are both positive.

$$Z + \Delta Z = (A + \Delta A)(B + \Delta B) = AB + B\Delta A + A\Delta B + \Delta A \Delta B$$

The errors in $A$ and $B$ are small in comparison with the values of $A$ and $B$ so we can neglect the quantity $\Delta A \Delta B$ as being insignificant. Then:

$$\Delta Z = B\Delta A + A\Delta B$$

Dividing through by $Z$ gives:

$$\frac{\Delta Z}{Z} = \frac{B\Delta A + A\Delta B}{Z} = \frac{B\Delta A + A\Delta B}{AB} = \frac{\Delta A}{A} + \frac{\Delta B}{B}$$

Thus, when we have the product of measured quantities, the worst possible fractional error in the calculated quantity is the sum of the fractional errors in the measured quantities. If we multiply the above equation by 100 then we can state it as the percentage error in $Z$ is equal to the sum of the percentage errors in the measured quantities. If we have the square of a measured quantity, then all we have is the quantity multiplied by itself and so the error in the squared quantity is just twice that in the measured quantity. If the quantity is cubed then the error is three times that in the measured quantity.

In a similar way we can show that if the calculated quantity is obtained by dividing one measured quantity by another, i.e. $Z = A/B$, then the worst possible fractional error in the calculated quantity is the sum of the fractional errors in the measured quantities or, if expressed in percentages, the percentage error in the calculated quantity is equal to the sum of the percentage errors in the measured quantities.

### Example

The resistance of a resistor is given by $V/I$. If the voltage $V$ has been measured as $2.1 \pm 0.2$ V and the current $I$ as $0.25 \pm 0.01$ A, what will be the error in the resistance value?

The percentage error in the voltage reading is $(0.2/2.1) \times 100 = 9.5\%$ and in the current reading is $(0.01/0.25) \times 100\% = 4.0\%$. Thus the percentage error in the resistance is $9.5 + 4.0 = 13.5\%$. Since the resistance is $V/I = 8.4 \, \Omega$ and 13.5% of 8.4 is 1.1, then the resistance is quoted as $8.4 \pm 1.1 \, \Omega$.

### Example

The cross-sectional area of a wire is given by $\frac{1}{4}\pi d^2$, where $d$ is the diameter. If the diameter has been measured as $1.2 \pm 0.1$ mm, what will be the error in the area value?

The percentage error in the diameter is $(0.1/1.2) \times 100 = 8.3\%$. Thus, because the diameter is squared, the percentage error is doubled. Hence the percentage error in the area is 16.6%. Since the area is $\frac{1}{4}\pi \times 1.2^2$ then the area is quoted as $1.13 \pm 0.19$ mm².

## 3.3 Tensile testing of materials

The following are a group of simple experiments that can be used to obtain information about the tensile properties of materials. *Safety note*: when doing experiments involving the stretching of wires, filaments, glass fibres or other materials, the specimen may fly up into your face when it breaks. When a taut wire snaps, a lot of stored elastic energy is suddenly released. *Safety spectacles should be worn.*

A simple appreciation of the behaviour of a range of materials can be obtained by just pulling the materials between your hands and feeling how they behave. Materials that might be pulled in this way include a rubber

Figure 3.7 *Pulling wires*

band, polyethylene from a food bag, nylon from a fishing line, copper wire, steel wire, nichrome wire, fuse wire, etc. In the case of the wires, the ends should be twisted round a pair of dowel rods (Figure 3.7). From such an experiment the materials can be classified in terms of strength, ductility, brittleness, elastic behaviour and plastic behaviour.

A force–extension graph for rubber can be obtained by hanging a rubber band (e.g. 74 mm by 3 mm by 1 mm band) over a clamp or other fixture, adding masses to a hanger suspended from it and measuring the extension with a ruler (Figure 3.8). A force–extension graph for nylon fishing line can be obtained in a similar way, the fishing line being tied to form a loop (e.g. about 75 cm long).

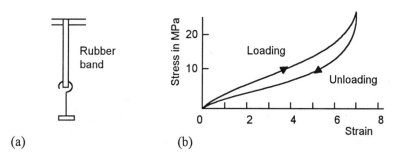

Figure 3.8 *Tensile test for rubber: (a) apparatus, (b) typical result*

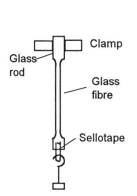

Figure 3.9 *Tensile test for glass*

A glass fibre for tensile testing can be obtained if the middle of a glass rod (e.g. about 3 mm diameter and 20 cm long) is softened in a Bunsen flame and drawn out into a fibre. The top of the glass rod can be clamped and a hanger attached to the lower end by means of Sellotape (Figure 3.9) or the lower end of the glass rod being formed into hook.

Figure 3.10 shows how the force–extension graph, and hence the tensile modulus of elasticity, can be determined for a metal wire (e.g. iron wire with a diameter of about 0.2 mm, copper wire about 0.3 mm diameter, steel wire about 0.08 mm diameter, all having lengths of about 2.0 m). The initial

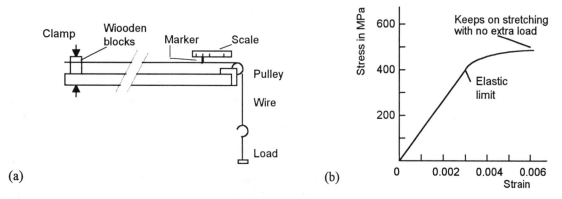

Figure 3.10 *Tensile test for wire: (a) apparatus, (b) example of result with copper wire*

diameter $d$ of the wire is measured using a micrometer screw gauge. The length $L$ of the wire from the clamped end to the marker (a strip of paper attached by Sellotape) is measured by using a rule, a small load being used to give a taut wire. Masses are then added to the hanger and the change in length $e$ from the initial position recorded. Hence data can be obtained to plot a graph of force ($F$) against extension ($e$). To obtain the tensile modulus of elasticity $E$, the gradient $F/e$ of the initial straight-line part of graph is determined. Then:

$$E = \frac{\text{stress}}{\text{strain}} = \frac{\left(F/\frac{1}{4}\pi d^2\right)}{e/L} = \frac{F}{e} \times \frac{4L}{\pi d^2}$$

### Example

How could the stress needed to break a glass fibre be determined?

Measure the diameter of the glass fibre by means of a micrometer screw gauge, then load the fibre (as in Figure 3.9) until it breaks. The breaking stress is the breaking load divided by the cross-sectional area of the fibre.

### Example

A length of wire was subject to a tensile test using the apparatus shown in Figure 3.10. The length of wire from the clamp to the marker was 2.0 m and the diameter of the wire 0.08 mm. Measurements were made of the movements of the marker as loads were added to the wire. A graph of load plotted against marker movement gave a straight-line graph through the origin with a slope of 0.051 kg/mm. What is the tensile modulus?

Using the equation derived above,

$$E = \frac{F}{e} \times \frac{4L}{\pi d^2} = 0.051 \times 1000 \times 9.8 \times \frac{4 \times 2}{\pi \times (0.08 \times 10^{-3})^2}$$

$$= 199 \times 10^9 \text{ Pa} = 199 \text{ GPa}$$

### Example

Estimate the error in the value of the modulus obtained in the previous example given the following information. The length of the wire was measured with a rule to an accuracy of $\pm 1$ mm. The diameter of the wire was measured a number of times at different places and with several readings in each place in case the wire was not perfectly circular and the results obtained were 0.081, 0.080, 0.079, 0.078, 0.080, 0.082, 0.081, 0.082, 0.079, 0.078 mm. The extensions of the wire were about 2 mm for each 0.1 kg load added. Because the extensions are small and difficult to measure accurately, the graph showed a scatter of points and the best straight line was used to give the gradient value. A number of other lines could have been drawn and it was considered that the degree of uncertainty in the value of the gradient was ±0.002 kg/mm.

The percentage error in the length is ±(1/200) × 100 = 0.5%. The mean value for the diameter is 8.00/10 = 0.080 mm. The deviations and the squares of the deviations are as shown in Table 3.2. The standard deviation is √(0.000 020/9) = 0.0015 mm. Hence the standard error is 0.0015/√(9 − 1) = 0.0005 mm and the percentage error in the diameter reading is ±(0.0005/0.080) × 100 = 0.6%. The percentage error in the square of the diameter will be twice that in the diameter and so is 1.2%. The percentage error in the value of the gradient, i.e. *F/e*, is given by ±(0.002/0.051) × 100 = 3.9%. The percentage error in the calculated tensile modulus value is the sum of the percentage errors in each of the measured quantities and so is 0.5 + 1.2 + 3.9 = 5.6%. Thus the possible error in the modulus value is ±199 × (5.6/100) = ±11 GPa. Hence the result of the experiment can be quoted as 199 ± 11 GPa.

Table 3.2 *Diameter readings*

| Diameter mm | Deviation mm | (Deviation)$^2$ mm$^2$ |
|---|---|---|
| 0.081 | +0.001 | 0.000 001 |
| 0.080 | 0 | 0 |
| 0.079 | −0.001 | 0.000 001 |
| 0.078 | −0.002 | 0.000 004 |
| 0.080 | 0 | 0 |
| 0.082 | +0.002 | 0.000 004 |
| 0.081 | +0.001 | 0.000 001 |
| 0.082 | +0.002 | 0.000 004 |
| 0.079 | −0.001 | 0.000 001 |
| 0.078 | −0.002 | 0.000 004 |

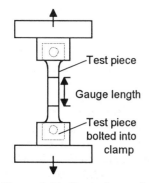

Figure 3.11 *Basic form of tensile test*

### 3.3.1 Industrial tensile testing

In an industrial tensile test, measurements are made of the force required to extend a standard test piece at a constant rate, the elongation of a specified gauge length of the test piece being measured by some forms of extensometer. Figure 3.11 shows the basic principle, a unidirectional force being applied to a test piece and the change in length between two marks, the gauge length, measured. British and European standards (BSEN 10002 Part I) state that the rate at which the stresses are applied should be between 2 and 10 MPa/s if the tensile modulus is less than 150 GPa and between 6 and 30 MPa/s if the tensile modulus is equal to or greater than 150 GPa.

In order to eliminate any variations in tensile test data due to differences in the shapes of test pieces, standard shapes and sizes are adopted. Test pieces are said to be *proportional test pieces* if the relationship between the gauge length $L_0$ and the cross-sectional area $A$ of the gauge length is

$$L_0 = k\sqrt{A}$$

European Standards specify the constant $k$ should have the value 5.65 and the gauge length should be 20 mm or greater. With circular cross-sections $A = \frac{1}{4}\pi d^2$ and thus to a reasonable approximation this value of $k$ gives

$$L_0 = 5d$$

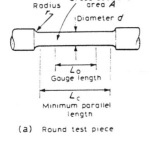

Radius — Cross-sectional area $A$ — Diameter $d$

$L_0$ Gauge length

$L_c$ Minimum parallel length

(a)  Round test piece

With circular cross-sectional areas that are too small for this value of $k$, a higher value may be used, preferably 11.3. When test pieces are proportional test pieces the same test results are given for the same test material when different size test pieces are used.

Figure 3.12 shows the standard size test pieces for proportional round and non-proportional flat samples of metals with Table 3.3 showing the standard dimensions that can be used. For the tensile test data for the same material to give essentially the same results, regardless of the length of the test piece used, it is vital that the standard dimensions are adhered to. An important feature of the dimensions is the radius given for the shoulders of the test pieces. Very small radii can cause localised stress concentrations that may result in the test piece failing prematurely.

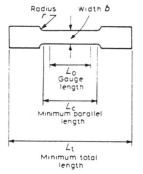

Radius   Width $b$

$L_0$ Gauge length

$L_c$ Minimum parallel length

$L_t$ Minimum total length

(b)  Flat test piece

Figure 3.12  *Standard tensile test pieces*

Table 3.3 *Dimensions of standard test pieces*

*Flat test pieces 0.1 to 0.3 mm thick*

| $b$ mm | $L_0$ mm | $L_c$ mm | $L_f$ mm |
|---|---|---|---|
| 20 | 80 | 120 | 140 |
| 12.5 | 50 | 75 | 87.5 |

*Round test pieces (proportional)*

| $d$ mm | $A$ mm$^2$ | $L_0$ mm | $L_c$ mm |
|---|---|---|---|
| 20 | 314.2 | 100 | 110 |
| 10 | 78.5 | 50 | 55 |
| 5 | 19.6 | 25 | 28 |

Note: $k = 5.85$

Tensile tests can likewise be used with plastic test pieces to obtain stress–strain data. The stress–strain properties of plastics are, however, much more dependent than metals on the rate at which the strain is applied. Thus, for example, the tensile test may indicate a yield stress of 62 MPa when the rate of elongation is 12.5 mm/min but 74 MPa when it is 50 mm/min. Also the form of the stress–strain graph may change with a ductile material at low strain rates becoming a brittle one at high strain rates. Figure 3.13 shows the general forms of stress–strain graphs for plastics at different strain rates.

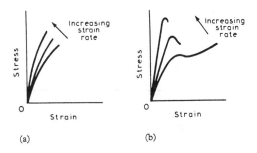

(a)                    (b)

Figure 3.13 *The effect of strain rate on (a) a brittle plastic, (b) a ductile plastic*

## 3.4 Bend tests

Some idea of the ductility of a material can be found by seeing how far you can bend a sheet of the material without it cracking and breaking. This is the basis of a simple test that is often quoted by suppliers of materials as a measure of ductility: the more ductile the material, the greater the angle it can bend through (Figure 3.14).

The test involves bending, in a specified way, a sample of the material through some angle and determining whether the material is unbroken and free from cracks after the bending. There are a number of ways that can be used to carry out such a test, BS 1639 lists the British standards. The simplest method is the mandrel test shown in Figure 3.15(a); this is suitable for medium and thin thickness sheet for angles of bend up to 120°. Figure 3.15(b) shows how the test can be conducted on a vee block; this is suitable for medium thickness sheet with bend angles up to 90°. Figure 3.15(c) shows the form of test possible for thin sheet with bend angles up to 90°, the material being bent on a block of soft material. Other methods can also be used, e.g. bending round a mandrel, free bending and pressure bending (see the British standard for more details).

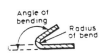

Figure 3.14 *The angle of bend*

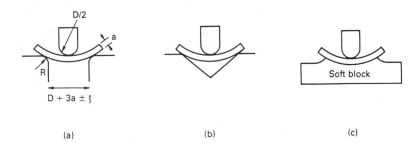

(a)                    (b)                    (c)

Figure 3.15 *Bend test: (a) mandrel test, (b) bending on a vee block, (c) bending on a block of soft material*

### 3.5 Impact tests

Impact tests are designed to simulate the response of a material to a high rate of loading and involve a test piece being struck a sudden blow. There are two main forms of test, the *Izod* and *Charpy* tests. Both tests involve the same type of measurement but differ in the form of the test pieces. Both involve a pendulum swinging down from a specified height $h_0$ to hit the test piece (Figure 3.16). The height $h$ to which the pendulum rises after striking and breaking the test piece is a measure of the energy used in the breaking. If no energy were used, the pendulum would swing up to the same height $h_0$ as that it started from, i.e. the potential energy $mgh_0$ at the top of the pendulum swing before and after the collision would be the same. The greater the energy used in the breaking, the lower the height to which the pendulum rises. If the pendulum swing up to a height $h$ after breaking the test piece, then the energy used to break it is $(mgh_0 - mgh)$.

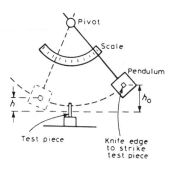

Figure 3.16 *The principle of the impact test*

### 3.5.1 Izod test pieces

With the Izod test, the energy absorbed in breaking a cantilevered test piece is measured, as illustrated by Figure 3.17. The test piece has a notch and the blow is struck on the same face as the notch and at a fixed height above it.

In the case of metals, the test pieces used are generally either 10 mm square or 11.4 mm in diameter if round. Figure 3.18(a) shows details of one form of the square test piece. With the 70 mm length, the notch is 28 mm from the top of the piece. If a longer length is used then more than one notch is used. With a length of 96 mm, there are two notches on opposite faces, one 28 mm from the top and the other twice that distance from the top. With a longer length test piece of 126 mm, there are three notches, on three of the faces. The first notch is 28 mm from the top, the second twice that distance and the third three times that distance from the top. In the case of plastics, the test pieces are either 12.7 mm square or 12.7 mm by 6.4 to 12.7 mm depending on the thickness of the material concerned. Figure

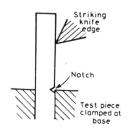

Figure 3.17 *Form of the Izod test (elevation view)*

3.18(b) shows details of such a test piece. With metals the pendulum strikes the test piece with a speed of between 3 and 4 m/s, with plastics a lower speed of 2.44 m/s is used.

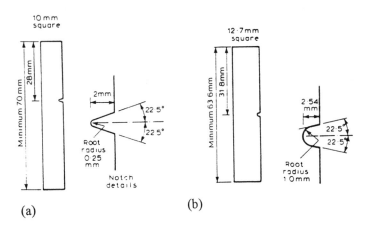

Figure 3.18 *Izod test piece for (a) a metal, (b) a plastic*

### 3.5.2 Charpy test pieces

With the Charpy test, the energy absorbed in breaking a test piece in the form of a beam is measured (Figure 3.19). The standard machine has the pendulum hitting the test piece with an energy of $300 \pm 10$ J. The test piece is supported at each end and notched at the midpoint between the two supports. The notch is on the face directly opposite to where the pendulum strikes the test piece. The British and European standard is BSEN 10045.

For metals, the test piece generally has a square cross-section of side 10 mm and length 55 mm with 40 mm between the supports. Figure 3.20 shows details of such a test piece and the forms of notch commonly used. With the V-notch, reduced width specimens of 7.5 mm and 5 mm can be used. For plastics, the test pieces may be unnotched or notched. A standard test piece is 120 mm long, 15 mm wide and 10 mm thick in the case of moulded plastics. With sheet plastics the width can be the thickness of the sheet with the thickness 15 mm or the width is between 5 and 10 mm with the thickness being 15 mm. The notch is U-shaped with a width of 2 mm and a radius of 0.2 mm at its base. For moulded plastics the depth below the notch is 6.7 mm, for the sheet plastics either 10 mm or two-thirds of the sheet thickness.

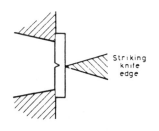

Figure 3.19 *Form of the Charpy test (plan view)*

### 3.5.3 Impact test results

In stating the results of impact tests it is vital that the form of test is specified. There is no reliable relationship between the values obtained by the two forms of test, so the values from one test cannot be compared with

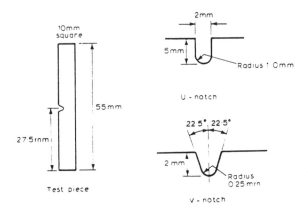

Figure 3.20  *Charpy test piece*

those from the other. In addition there is no reliable relationship between the impact energies given for breaking test pieces of different sizes or different notches with the same test method. The impact energy is influenced by such factors as the temperature, the speed of impact, any degree of directionality in the properties of the material from which the test piece was cut, and the thickness of the test piece. For both the Izod and Charpy tests, the impact strengths for metals are expressed in the form of the energy absorbed, i.e. as 30 J. For plastics, with the Izod test the results are expressed as the energy absorbed in breaking the test piece divided by the width of notch, and with the Charpy test as the energy absorbed divided by either the cross-sectional area of the specimen for unnotched test pieces or by the cross-sectional area behind the notch for notched test pieces, e.g. $2 \text{ kJ/m}^2$.

When a material is stretched, energy is stored in the material. Think of stretching a spring or a rubber band. When the stretching force is released the material springs back and the energy is released. However, if the material suffers a permanent deformation, then all the energy is not released. The greater the amount of such plastic deformation, the greater the amount of energy not released. Thus when a ductile material is broken, more energy is 'lost'. The fracture of materials can be classified roughly as either brittle or ductile fracture. With brittle fracture there is little plastic deformation prior to fracture and so little energy is required to break the test piece. With ductile fracture, the fracture is preceded by a considerable amount of plastic deformation and so more energy is required to break the test piece. Thus the impact test can be used to give information about the type of fracture that occurs. For example, Figure 3.21 shows the effect of temperature on the Charpy V-notch impact energies obtained for test pieces of a 0.2% carbon steel. Above about 0°C, the material gives ductile failures, below that temperature brittle failures. Thus at low temperatures the steel can be easily shattered by impact.

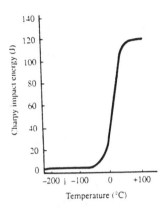

Figure 3.21 *Effect of temperature on the Charpy V-notch impact energies for a 0.2% carbon steel*

The appearance of the fractured surfaces after an impact test also gives information about the type of fracture that has occurred. With a brittle fracture of metals, the surfaces are crystalline in appearance. With a ductile fracture, the surfaces are rough and fibrous. Also with ductile failure, there is a significant reduction in the cross-sectional area of the test piece, but with brittle fracture, there is virtually no such change. With plastics, a brittle failure gives fracture surfaces that are smooth and glassy or somewhat splintered; with a ductile failure the surfaces often have a whitened appearance. Also, with plastics, the change in cross-sectional area can be considerable with a ductile failure but negligible with brittle failure.

### Example

A sample of unplasticised PVC has an impact strength of 3 kJ/m$^2$ at 20°C and 10 kJ/m$^2$ at 40°C. Is the material becoming more or less brittle as the temperature is increased?

Because there is an increase in the impact energy, the material is becoming more ductile.

## 3.6 Creep and fatigue

There are many situations where materials are subject to loads for protracted periods of time and the resulting strain does not remain constant but slowly changes with time, even though the applied stress remains constant. This effect is called *creep*. Figure 3.22 show a simple experiment that can be used to study creep. A strip of metal, or plastic, is clamped at one end to form a cantilever. A weight is attached to the free end, so that the strip bends initially through an angle of about 30°. The deflection of the

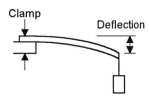

Figure 3.22 *Creep experiment*

free end is measured initially and then again after, say, 30 seconds, 1 minute, 2 minutes, 5 minutes, 10 minutes, 20 minutes, 40 minutes, etc. A graph can then be plotted showing how the deflection, a measure of the strain in the material, varies with time.

Creep can be quite significant for plastics at room temperature and for metals at temperatures that approach their melting points. Industrial creep measurements with metals are thus generally carried out at elevated temperatures. Figure 3.23 shows the type of apparatus used and the type of result obtained.

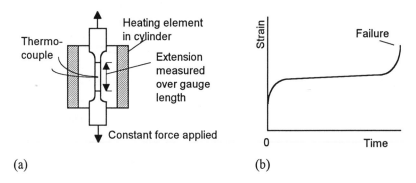

(a)                                                                            (b)

Figure 3.23 *Creep measurement: (a) apparatus, (b) typical result*

Another situation where the results may vary with time is *fatigue*. This is when a component in service is subject to situations where there is a fluctuating stress. This may be repeatedly stressed and unstressed, or alternately in tension and compression. A simple experiment to investigage fatigue is to take strips of material, metals and polymers, and flex them back and forth by hand. You might just try a straightened-out paper clip. Many plastics will show little effect and remain unbroken after many such flexures. Metals, however, are very likely to break.

Industrial fatigue tests involve using machines to carry out such fluctuating stresses, the type of fluctuation used being chosen to mirror the type of stress changes the material will experiment in service.

## 3.7 Hardness tests

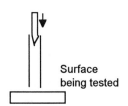

Figure 3.24 *Simple hardness test*

The hardness of a material may be specified in terms of some standard test involving indenting or scratching the surface; the harder a material, the more difficult it is to make an indentation or a scratch. A simple experiment (Figure 3.24) that can be used to compare hardness is to take a centre punch and drop it vertically through, say, 30 cm on to the surface being tested (so that it drops vertically, you might drop it down a tube). The size of the indentation can then be taken as a measure of the hardness. The measure of the size might be the diameter of the indentation.

### 3.7.1 Industrial tests

The most common form of hardness tests for metals involves standard indentors being pressed into the surface of the material concerned. Measurements associated with the size of the indentation are then taken as a measure of the hardness of the surface. The Brinell test, the Vickers test and the Rockwell test are the main forms of such tests. There is no absolute scale for hardness, each test having its own scale. Though some relationships exist between results on one scale and those on another, care has to be taken in making comparisons because the different types of test are measuring different things.

With the *Brinell test*, a hardened steel ball is pressed for a time of 10 to 15 s into the surface of the material by a standard force (Figure 3.25). After the load and ball have been removed, the diameter of the indentation is measured. The Brinell hardness number (signified by HB) is obtained by dividing the size of the force applied by the surface area of the spherical indentation.

$$\text{Brinell hardness number} = \frac{\text{applied force}}{\text{area of indentation}}$$

The units used for the area are mm$^2$ and for the force kgf (1 kgf = 9.8 N, the gravitational force exerted by 1 kg). The area can be obtained, from the measured diameter of the indentation and ball diameter, either by calculation or the use of tables.

$$\text{Area} = \tfrac{1}{2}\pi D\left[ D - \sqrt{(D^2 - d^2)}\, \right]$$

where $D$ is the diameter of the ball and $d$ that of the indentation. The diameter $D$ of the ball used and the size of the applied force $F$ are chosen, for the British Standard, to give $F/D^2$ values of 1, 5, 10 or 30 with the diameters of the balls being 1, 2, 5 or 10 mm. In principle, the same value of $F/D^2$ should give the same hardness value, regardless of the diameter of the ball used.

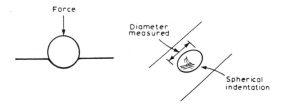

Figure 3.25 *The basis of the Brinell hardness test*

The Brinell test cannot bẹ used with very soft or very hard materials. In the one case, the indentation becomes equal to the diameter of the ball, and in the other, there is either no or little indentation on which measurements can be based. The thickness of the material being tested should be at least ten times the depth of the indentation if the results are not to be affected by the thickness of the material.

The *Vickers hardness test* involves a diamond indenter being pressed under load for 10 to 15 s into the surface of the material under test (Figure 3.26). The result is a square-shaped impression. After the load and indenter are removed, the diagonals $d$ of the indentation are measured. The Vickers hardness number (HV) is obtained by dividing the size of the force, in units of kgf, applied by the surface area, in mm$^2$, of the indentation.

$$\text{Vickers hardness} = \frac{\text{applied force}}{\text{area of indentation}}$$

The surface area can be calculated, the indentation being assumed to be a right pyramid with a square base and an apex angle $\theta$ of 136°, or obtained by using tables and the diagonal values.

$$\text{Area} = \frac{d^2}{2 \sin \theta/2} = \frac{d^2}{1.854}$$

The Vickers test has the advantage over the Brinell test of the increased accuracy that is possible in determining the diagonals of a square as opposed to the diameter of a spherical mark. Otherwise it has the same limitations as the Brinell test.

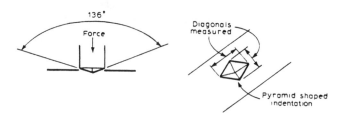

Figure 3.26 *The basis of the Vickers hardness test*

The *Rockwell hardness test* differs from the Brinell and Vickers tests in not obtaining a value for the hardness in terms of the area of an indentation but using the depth of indentation. The test uses either a diamond cone or a hardened steel ball as the indenter (Figure 3.27). A force of 90.8 N (10 kgf) is applied to press the indenter into contact with the surface. An additional force is then applied and causes an increase in depth of indenter penetration

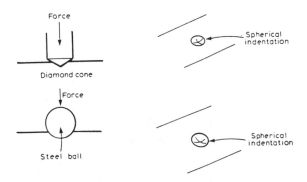

Figure 3.27 *The basis of the Rockwell hardness test*

into the material. The additional force is then removed and there is some reduction in the depth of the indenter due to the deformation of the material not being entirely plastic. The difference in the final depth of the indenter and the initial depth, before the additional force was applied, is determined. This is the permanent increase in penetration $e$ due to the additional force. The Rockwell hardness number (HR) is then given by:

Rockwell hardness number = $E - e$

where $E$ is a constant determined by the form of the indenter. For the diamond cone indenter $E$ is 100, for the steel ball 130.

There are a number of Rockwell scales, the scale being determined by the indenter and the additional force used. Table 3.3 indicates the scales and the types of materials for which each is typically used. Note that the diameter of the balls arise from standard sizes in inches, 1.588 mm being 1/16 in, 3.175 mm being 1/8 in, 6.350 mm being 1/4 in, and 12.70 mm being 1/2 in. In any reference to the results of a Rockwell test, the scale letter must be quoted. For metals, the B and C scales are probably the most commonly used ones.

The Brinell and Vickers tests both involve measurements of the surface area of indentations, the forms of the indenters used being different. The Rockwell test involves measurements of the depth of penetration of indenters. Thus the various tests are concerned with different measurements as an indication of hardness. Consequently the values given by the different methods differ for the same material. Figure 3.28 shows the general range of hardness values for different types of material when measured by the Vickers, Brinell and Rockwell test methods.

There is an approximate relationship between hardness values and tensile strengths. Thus for annealed steels, the tensile strength in MPa is about 3.54 times the Brinell hardness value, and for quenched and tempered steels 3.24 times the Brinell hardness value. For brass the factor is about 5.6 and for aluminium alloys about 4.2.

Table 3.3  *Rockwell hardness scales*

| Scale | Indenter | Load kg | Typical applications |
|---|---|---|---|
| A | Diamond | 60 | Extremely hard materials, e.g. tool steels |
| B | Ball 1.588 mm dia. | 100 | Soft materials, e.g. Cu alloys |
| C | Diamond | 150 | Hard materials, e.g. steels |
| D | Diamond | 100 | Medium case hardened metals |
| E | Ball 3.175 mm dia. | 100 | Soft materials, e.g. Al alloys |
| F | Ball 1.588 mm dia. | 60 | As E, the smaller ball used where inhomogeneities exist |
| G | Ball 1.588 mm dia. | 150 | Malleable irons, bronzes |
| H | Ball 3.175 mm dia. | 60 | Al, Pb and Zn |
| K | Ball 3.175 mm dia. | 150 | Al and Mg alloys |
| L | Ball 6.350 mm dia. | 60 | Plastics |
| M | Ball 6.350 mm dia. | 100 | Plastics |
| P | Ball 6.350 mm dia. | 150 | Plastics |
| R | Ball 12.70 mm dia. | 60 | Plastics |
| S | Ball 12.70 mm dia. | 100 | Plastics |
| V | Ball 12.70 mm dia. | 150 | Plastics |

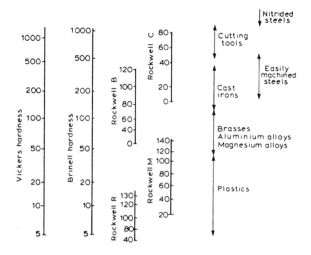

*Figure 3.28 Hardness values*

## 3.8 Photoelastic stress analysis

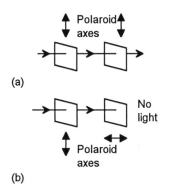

(a)

(b)

Figure 3.29 *Light passing through Polaroids*

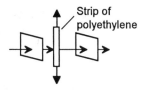

Figure 3.30 *Photoelastic stress analysis*

The applied forces that will cause a strip of material to break are very much dependent on whether the material has any initial cracks, notches, holes or other 'defects' which result in a stress concentration. Think of how much easier it is to tear a piece of paper if it already has perforations or some initial tear. A way that can be used to illustrate this is *photoelastic stress analysis*.

When ordinary light passes through a piece of Polaroid, only the light waves vibrating in a particular plane are transmitted. If this light is then passed through a second piece of Polaroid, then if the axis of this second Polaroid is the same as that of the first one, the light passes straight through because its plane of vibration is still in the right plane (Figure 3.29(a)). However, if the second Polaroid has its axis at right angles to that of the first Polaroid, no light passes through it (Figure 3.29(b)). This is because the plane of vibration of the light transmitted through the first Polaroid is in the wrong direction for transmission through the second Polaroid.

Certain transparent materials have the property that when they are stressed and viewed between crossed pieces of Polaroid, they show a pattern of coloured fringes. Each coloured fringe corresponds to a particular level of stress. They are thus similar to the contour lines on a map, but instead of mapping height levels, they map stress levels. With a map, where there is a steep hill, we get contour lines close together. Thus where there is a high stress concentration, we get coloured fringes close together. Thus stress concentrations can be readily determined.

A similar experiment can demonstrate such fringes. With a pair of crossed Polaroids, a strip of heavy gauge polyethylene should be placed between them (Figure 3.30). The fringe pattern resulting when the polyethylene strip is pulled can be compared with that which occurs when the strip has a small slit cut in one edge. The slit gives rise to a greater stress concentration. Another effect that can be examined is that of a strip with a central hole. Again there is a stress concentration.

## 3.9 Electrical tests

The measurement of the electrical resistivity or conductivity of a material requires a measurement of the resistance of a strip or block of the material. In the case of metals, the resistivity is very low and so the resistance can be low. For example, the resistance of a 1 m length of copper wire with a diameter of 1 mm is about 0.03 $\Omega$ at 20°C. Such a resistance is not easy to measure, since the means by which it is connected to the measurement system can have resistances of the same order of size or even larger. A smaller gauge wire of 0.1 mm gives a resistance of about 2.1 $\Omega$ and is easier to measure. An even smaller diameter would give an even higher resistance and be even easier to measure.

A simple and quick method for the measurement of resistance is the *ammeter–voltmeter method*. An ammeter is connected in series with the length of wire and a voltmeter in parallel with it (Figure 3.31). Readings are taken of the current and voltage for different input voltages to the circuit. A graph of the voltage across the resistance wire plotted against the current through it will give, in the case of metals, a straight-line graph. The gradient of the graph is the resistance $R$ of the length of wire. To obtain the resistivity, the length of the wire and its diameter have to be measured. The diameter should be measured at a number of points and an average value obtained.

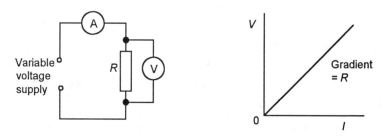

Figure 3.31  *Ammeter–voltmeter method*

The accuracy of the ammeter–voltmeter method of resistance measurement depends on the accuracy of the meters used and the resistance of the voltmeter used. The ammeter measures not only the current through the resistance wire but also the current through the voltmeter. The higher the resistance of the voltmeter relative to that of the resistance being measured, the smaller the fraction of the current that passes through the voltmeter and so the closer the reading of the ammeter is to the current through the resistance. For the measurement of resistances of a few ohms with a voltmeter with resistance of thousands of ohms, the effect on the accuracy is likely to be considerably smaller than the limitations imposed by the accuracy with which the ammeter and voltmeter can be read. With ammeters and voltmeters having pointers moving across scales, the accuracy with which readings can be made is likely to be of the order of a few per cent. For example, with a voltmeter with a scale of 0 to 1 V and scale divisions of 0.1 V we might consider that we can estimate the position of the pointer setting to within ±0.2 of a scale division. This is an accuracy of ±0.02 V. In a full-scale reading of 1 V this would represent an accuracy of ±2%.

The *Wheatstone bridge* gives a more accurate method of measuring resistance. Figure 3.32 shows the basic form of such a bridge. The resistances $R_1$, $R_2$, $R_3$ and $R_4$ in the arms of the bridge are adjusted so that there is no current through the galvanometer. In such a condition the bridge is said to be balanced.

In order to develop the theory of the bridge we have to consider the potential divider circuit shown in Figure 3.33(a). Because the two resistors are in series, the current through each of them will be the same. Thus the potential difference across $R_3$ is $I_2R_3$ and that across $R_4$ is $I_2R_4$. The supply voltage $V_s$ is thus divided between the two resistors according to the value of their resistances. Likewise, for the circuit shown in Figure 3.33(b), the current through $R_1$ and $R_2$ will be the same and so the potential difference across $R_1$ is $I_1R_1$ and that across $R_2$ is $I_1R_2$. The supply voltage $V_s$ is thus

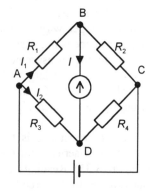

Figure 3.32  *Wheatstone bridge*

divided between the two resistors according to their resistances. If we combine these two circuits, we have the Wheatstone bridge circuit.

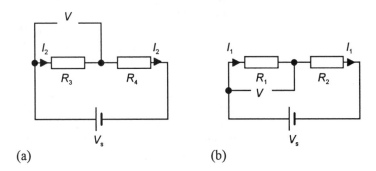

(a)                                      (b)

Figure 3.33  *Potential divider circuits*

When there is no current through the galvanometer (Figure 3.32), then there must be no potential difference across it. Points B and D must be at the same potential. Thus the potential difference across $R_1$ must be the same as that across $R_3$, i.e.

$$I_1 R_1 = I_2 R_3$$

and so:

$$\frac{R_1}{R_3} = \frac{I_2}{I_1}$$

It also means that the potential difference across $R_2$ is the same as that across $R_4$. Thus, since there is no current through the galvanometer, the current through $R_1$ must be the same as that through $R_2$ and the current through $R_3$ the same as that through $R_4$. Hence:

$$I_1 R_2 = I_2 R_4$$

and so:

$$\frac{R_2}{R_4} = \frac{I_2}{I_1}$$

Thus we must have:

$$\frac{R_1}{R_3} = \frac{R_2}{R_4}$$

This balance condition is independent of the supply voltage, depending only on the resistances in the four arms of the bridge. The galvanometer only has to determine whether there is a current and so the result is not affected by the calibration of the instrument. It is termed a *null method* since zero current is being looked for. The Wheatstone bridge is thus capable of high precision and is widely used for the measurement of resistances in the range $1 \Omega$ to $1 M\Omega$.

Figure 3.34 shows a version of the Wheatstone bridge called the *metre bridge*. The resistance being determined, e.g. a length of wire, is $R_1$ and an accurately known resistance is used for $R_2$, e.g. a resistance box. The resistances $R_3$ and $R_4$ are provided by a length of uniform resistance wire of length $L$, typically one metre. The resistance per unit length $r$ is thus the same at all points along the wire. The jockey, a knife-edged movable contact, is moved to a point along the wire that results in zero current through the galvanometer. If this occurs with the jockey a distance $x$ from one end, then we have $R_3 = xr$ and $R_4 = (L - x)r$. Hence, using the above balance equation,

$$R_1 = \frac{R_3}{R_4} \times R_2 = \frac{xr}{(L-x)r} \times R_2 = \frac{x}{L-x} \times R_2$$

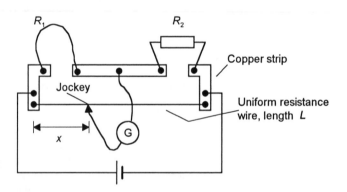

Figure 3.34 *The metre bridge*

The greatest accuracy is obtained when the balance point is near the centre of $L$, both $x$ and $(L - x)$ can then be determined to a reasonable degree of accuracy.

### Example

The ammeter–voltmeter method is used to determine the resistance of a length of nichrome wire. The length of the wire is measured as 200 mm with an accuracy of ±1 mm. The diameter is measured a number of times and gives an average value of 0.122 mm with an accuracy of ±0.002 mm. The resistance of the wire is determined from a single measurement of the voltage and current, the current reading being 110 mA when the voltage reading is 2.0 V. The reading accuracy of the ammeter is estimated as being ±2 mA and that of the voltmeter ±0.1 V. Determine the resistivity and its accuracy.

The resistivity ρ is given by

$$\rho = \frac{V\pi d^2}{4IL} = \frac{2.0 \times \pi \times (0.122 \times 10^{-3})^2}{4 \times 0.110 \times 0.200}$$

Hence the resistivity is $1.06 \times 10^{-6}$ Ω m.

The percentage accuracy of the length is $(1/200) \times 100 = 0.5\%$. The percentage accuracy of the diameter is $(0.002/0.122) \times 100 = 1.6\%$. The percentage accuracy of the square of the diameter will be twice that of the diameter, namely 3.2%. The percentage accuracy of the current is $(2/110) \times 100 = 1.8\%$. The percentage accuracy of the voltage is $(0.1/2.0) \times 100 = 5.0\%$. Adding all the percentage accuracies gives $0.5 + 3.2 + 1.8 + 5.0 = 10.5\%$. This therefore means an accuracy of $(10.5/100) \times 1.06 \times 10^{-6} = 0.11 \times 10^{-6}$ Ω m. Hence the resistivity result can be quoted as $1.06 \times 10^{-6} \pm 0.11 \times 10^{-6}$ Ω m.

If we had required the conductivity, i.e. 1/resistivity, then the percentage accuracy would have been the same and the result quoted as $9.43 \times 10^5 \pm 0.99 \times 10^5$ S/m.

### Example

A 20 cm length of a copper–manganese alloy wire is placed in one gap of a metre bridge and a standard resistance of 10 Ω in the other gap. Balance is achieved when the jockey is 41.1 cm from the end of the wire, length 100.0 cm, nearest the unknown resistance end. Determine the resistance of the wire.

Using the equation developed above,

$$R_1 = \frac{x}{L-x} \times R_2 = \frac{41.1}{100.0 - 41.1} \times 10 = 6.98 \text{ Ω}$$

### Example

Determine the accuracy of the measurement in the previous example if the balance point can be determined to ±0.1 mm of wire and the standard resistance has an accuracy of ±0.05 Ω.

Assuming that there are no errors associated with the end contacts between the wire and the copper strips, then the percentage accuracy in $x$ is $\pm(1/41.1) \times 100 = \pm2.4\%$ and the percentage accuracy in $(L - x)$ is $\pm(1/58.9) \times 100 = 1.7\%$. The percentage accuracy of the standard resistor is $\pm(0.05/10) \times 100 = 0.5\%$. Thus the overall percentage accuracy is $2.4 + 1.7 + 0.5 = 4.7\%$ and the result can be quoted as $6.98 \pm 0.33$ Ω.

### 3.9.1 Industrial methods for the measurement of resistivity

The British standard for resistivity measurements with metals is BS 5714. For routine resistance measurements the method used is likely to be a resistance bridge capable of an accuracy of at least ±0.30%. In addition to measuring the resistance, the length and cross-sectional area of the test piece are required. The area can be obtained by direct measurement, however, an alternative method that is often used is to weigh the test piece and calculate the area from a knowledge of the density and length, the area being mass/(density × length).

Since resistivity changes with temperature, it is important that the temperature $t$ at which a measurement is made is noted. The following equation is used to correct the result to the reference temperature $t_0$ at which the result is required:

$$\rho_{t_0} = \frac{\rho_t}{1 + (\alpha + \gamma)(t - t_0)}$$

where $\alpha$ is the temperature coefficient of resistance at the reference temperature and $\gamma$ the coefficient of linear expansion.

Materials such as plastics or ceramics have very high resistivities. This can present the problem in that the surface layers, perhaps as a result of the absorption of moisture, might have a significantly lower resistivity than the bulk of the material and so the value indicated by the measurement is not that of the bulk material. Polymers also present the problem that when a voltage is applied across a sample, the current through the material slowly decreases with time. Thus resistivity measurements need to have a time quoted with them, e.g. the value 60 s after the application of a voltage.

## 3.10 Thermal property tests

The following indicate methods that can be used in a school/college laboratory for the measurement of specific heat capacities and, in principle, are similar to methods that are used in industrial laboratories.

Figure 3.35 shows how the specific heat capacity of a cylindrical block of metal can be measured. The block of metal has been drilled with holes for an electrical heater (12 V, 2 to 4 A) and a thermometer. The block is weighed to give the mass $m$. Then the block is lagged with a poor thermal conductor, e.g. expanded polystyrene. The initial temperature $\theta_1$ of the block is recorded. The voltage is then switched on as a stop clock is started, the voltmeter and ammeter readings $V$ and $I$ being recorded. When the temperature has risen by about 10°C, the voltage is switched off and the time $t$ noted. The highest reading $\theta_2$ given by the thermometer is noted. If we assume that no energy losses occur, then the electrical energy supplied by the heater must equal the heat received by the block. Thus

$$IVt = mc(\theta_2 - \theta_1)$$

where $c$ is the specific heat capacity of the metal block. The small amount of heat received by the thermometer and the heater has been ignored.

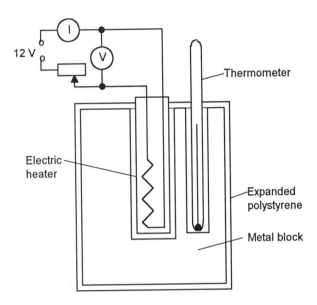

Figure 3.35 *Measurement of specific heat capacity of a metal*

Figure 3.36 shows the comparable apparatus that can be used for the determination of the specific heat capacity of a liquid. The inner calorimeter contains a known mass of the liquid. In order to reduce heat losses, the inner calorimeter is contained in a closed outer vessel so that it ends up being surrounded by still air, a good thermal insulator. The procedure is similar to that used for the solid. The liquid has to be continuously stirred during the heating in order to ensure that the thermometer accurately represents the temperature of the liquid as a whole. If we assume that no energy losses occur, then the electrical energy supplied by the heater must equal the heat received by the liquid plus the heat received by the inner calorimeter and stirrer. Thus

$$IVt = mc(\theta_2 - \theta_1) + m_c c_c(\theta_2 - \theta_1)$$

where $m_c$ is the mass of the inner calorimeter and stirrer and $c_c$ the specific heat capacity.

In the above discussion of measurements to obtain the specific heat capacities of solids and liquids, it has been assumed that no heat losses occur. Precautions that are taken to minimise heat losses include: surrounding the sample by an outer container or a jacket of a poor heat conductor to reduce convection and conduction losses, polishing the outer wall of the calorimeter to reduce radiation losses, and perhaps precooling the sample to below room temperature by, say, 5°C and then heating until it is 5°C above room temperature so that the heat gained by the sample when it is below room temperature cancels out the heat lost when it is above room temperature.

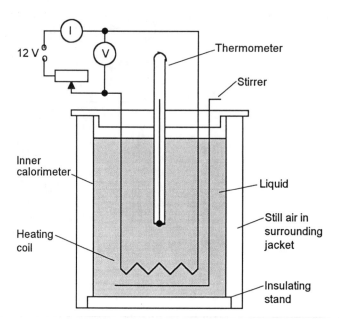

Figure 3.36 *Measurement of specific heat capacity of a liquid*

**Example**

In an experiment to determine the specific heat capacity of a liquid, 180 g of it was contained in a calorimeter of mass 100 g. For 100 s a voltage of 10.0 V was applied to an electrical heater immersed in the stirred liquid, the current being 4.0 A. This raised the temperature of the liquid from 5°C below room temperature to 5°C above it. If the calorimeter has a specific heat capacity of 400 J kg$^{-1}$ K$^{-1}$, what is the specific heat of the liquid?

Using the equation derived above, namely,

$$IVt = mc(\theta_2 - \theta_1) + m_c c_c(\theta_2 - \theta_1)$$

then

$$4.0 \times 10.0 \times 100 = 0.180 \times c \times 10 + 0.100 \times 400 \times 10$$

Thus $c = 2000$ J kg$^{-1}$ K$^{-1}$.

## 3.11 Optical property tests

Measurements of the refractive index of transparent materials are necessary if the materials are to be used in the design of optical instruments. However, refractive index measurements are more widely used in chemical laboratories as a means of determining the concentration of organic substances in solvents. In the main, such measurements tend to involve critical angle measurements.

The following is a critical angle experiment that can be used in a school/college laboratory for the determination of the refractive index of a liquid. Two parallel-sided glass plates are cemented together so as to enclose a thin layer of air (this might be a pair of microscope slides separated by a frame of cardboard and then cemented together). The plates, termed an *air cell*, are then immersed in the liquid under test, this being contained in a parallel-sided glass tank (Figure 3.37). A beam of light is then directed through the tank. When the cell is at right angles to the beam, the light passes straight through and can be seen on the far side of the tank. At other angles we can have the situation shown in Figure 3.38. However, when the cell is rotated, a position is reached when the beam is at such an angle that it becomes total internally reflected at the glass-to-air interface. When this occurs, there is no beam transmitted through the cell. The angle through which the cell has to be rotated for this cut-off of the transmitted light to occur is then measured by a pointer attached to the cell and moving over an angular scale. The angle is usually measured for the cell being rotated from the straight-through position in a clockwise direction and in an anticlockwise direction and the average value taken.

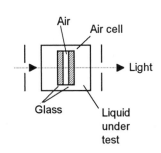

Figure 3.37 *The air cell method*

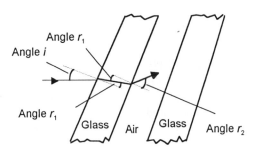

Figure 3.38 *The air cell*

With the situation shown in Figure 3.38, at the liquid to glass interface we have:

$$_l n_g = \frac{\sin i}{\sin r_1}$$

where $_l n_g$ is the air-to-glass refractive index. At the glass-to-air interface we have:

$$_g n_a = \frac{\sin r_1}{\sin r_2}$$

Thus:

$$_i n_g \times {}_g n_a = \frac{\sin i}{\sin r_1} \times \frac{\sin r_1}{\sin r_2} = \frac{\sin i}{\sin r_2}$$

But $\sin i_1 / \sin r_2$ is what we would have if the light had passed directly from liquid to air. Thus:

$$_i n_a = \frac{\sin i}{\sin r_2}$$

When the light is at the critical angle to the glass–air interface, i.e. $r_2 = 90°$, then no light is transmitted through the cell. Then we have:

$$_i n_a = \frac{\sin i}{\sin 90°} = \sin i$$

Since the refractive index going from air to the liquid $_a n_i$ is just the reciprocal of that going from liquid to air,

$$_a n_i = \frac{1}{\sin i}$$

Thus a measurement of the angle of incidence for cut-off of the transmitted light gives the refractive index of the liquid.

### Example

The refractive index of a liquid is being determined by the air cell method. The angle at which cut-off of the beam of light occurs is measured as 42°. The accuracy with which this angle can be determined is considered to be ±1°. Determine the refractive index of the liquid and the accuracy of the value.

The refractive index is given by the equation derived above, namely:

$$_a n_i = \frac{1}{\sin i} = \frac{1}{\sin 42°} = 1.494$$

With an error of +1° the refractive index value would be:

$$_a n_i = \frac{1}{\sin i} = \frac{1}{\sin 43°} = 1.467$$

With an error of −1° the refractive index value would be:

$$_a n_i = \frac{1}{\sin i} = \frac{1}{\sin 41°} = 1.524$$

Thus the refractive index is $1.49 \pm 0.03$.

### 3.11.1 Industrial method for the measurement of refractive index

An instrument used for the measurement of refractive index is termed a *refractometer*. A commonly used refractometer for the measurement of the refractive index of liquids is the *Abbé refractometer*. Figure 3.39 shows the basic principle. The liquid under test is placed between two glass prisms. Light from a diffuse light source is incident on prism A at a wide range of angles. The light is thus refracted through the prism at a range of angles and is then incident on the liquid layer. Some of the light will be incident at angles less than the critical angle and some greater than it. Thus some of the light will pass through into prism B and enter a telescope and some will be totally internally reflected at the glass–liquid interface and not enter the telescope. As a consequence, on moving the telescope round, a sharp division between light and dark will be found at the critical angle. With the cross-wires on this light–dark boundary, the refractive index can be read from the angular position of the telescope against a calibrated scale.

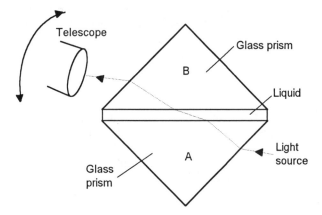

Figure 3.39 *Principle of the Abbé refractometer*

## 3.12 Chemical property tests

Materials in service can be affected by their environment and properties, such as the mechanical properties of strength and toughness, changed. Measurements of these properties over a period of time can give an indication of the interaction of the environment with a material.

Metallic materials corrode in moist air, with some metals corroding at a faster rate than others. Metals exposed to corrosive environments are often protected by being coated with a material such as paint. Tests are then used to investigate the weathering characteristics of the painted material. An accelerated weathering process is often used with exposure to radiation from an electric arc to simulate exposure to the sun and intermittent exposure to a spray of water to simulate rain. BS 6917 gives details of corrosion testing in artificial atmospheres, indicating the requirements for specimens, apparatus and procedures. BS 3900: Part G gives details of environmental tests on paint films.

The use of metals at high temperatures is often restricted by surface attack or scaling which gradually reduces the cross-sectional area and hence the stress-bearing ability of the item. The build-up of oxide layers at high temperatures is very much influenced by the environment, e.g. metal pipes exposed to superheated steam or hot gases from furnaces. Materials are tested by exposing them to such situations and measuring the reduction in the metal thickness as a consequence of the corrosive attack.

Plastic materials may dissolve in some liquids or absorb sufficient of the liquid to have their properties changed. When absorption occurs, the plastic becomes permeable to the liquid, i.e. liquid can leak through it. This permeability is of vital concern if the plastic is being considered for used as a container for liquids, e.g. a Coca-Cola bottle (see Section 1.2). Plastics are not generally subject to corrosion in the same way as metals but they can be adversely affected by weathering, i.e. exposure to light, heat, rain, sun. This can show itself as a fading of the colour of the plastic and/or a loss of flexibility. Tests are used to determine the weathering resistance of plastics with them being subject to an accelerated weathering process involving exposure to radiation from an electric arc, with intermittent exposure to a spray of water to simulate rain. The tests tend to be comparative ones with standard colours/materials being simultaneously exposed and performances compared.

## 3.12 Magnetic property tests

Figure 3.40 shows an experiment that can be used to determine the hysteresis loop of magnetic materials. A Hall probe is used to monitor the flux density in the material produced when the magnetising field is cycled by an alternating current passing through a coil placed round the sample of the material.

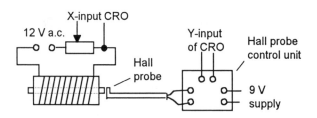

Figure 3.40 *Determination of hysteresis loop*

Possible materials that can be tried are rods of soft iron, steel, nickel, a hacksaw blade, a ferrite aerial rod, etc. The magnetising field $B_0$ is monitored by the X-deflection of a cathode ray oscilloscope, the flux density in the material $B$ by the Y-deflection plates. The result is the hysteresis loop automatically plotted on the screen of the oscilloscope.

**Problems**    *Questions 1 to 10 have four answer options: A, B, C and D. Choose the correct answer from the answer options.*

Questions 1 to 3 relate to the following information:

Repeated measurements of the force necessary to break tensile test pieces cut from a sample of the same material gave the following results:

102, 105, 103, 108, 105 N

1  The average value, to one decimal place, of the force necessary to break the material is:

A  104.6 N
B  105.8 N
C  106.0 N
D  108.0 N

2  The standard deviation, to one decimal place, of the sample is:

A  0.6 N
B  2.1 N
C  4.6 N
D  5.8 N

3  The standard error, to one significant figure, of the mean is:

A  0.4 N
B  0.5 N
C  1 N
D  2 N

4  A sample of $n$ measurements of a variable has a standard deviation of $\sigma$. The 95.5% confidence interval for the mean is given, to a reasonable approximation, by the mean plus or minus:

A  $2\sigma$
B  $\dfrac{2\sigma}{n}$
C  $\dfrac{2\sigma}{\sqrt{n}}$
D  $\sqrt{\dfrac{2\sigma}{n}}$

5  A student is setting up an experiment to determine Young's modulus for a steel in the form of a wire. The experiment involves stretching a length of the wire by measured loads and determining the extensions produced.

Decide whether each of these statements is TRUE (T) or FALSE (F).

The accuracy of the experiment will be increased if:
(i)   A long length of wire is used because the extensions produced will be larger than with a smaller length.
(ii) A large diameter wire is used because it will stretch more for a given load.

A   (i) T   (ii) T
B   (i) T   (ii) F
C   (i) F   (ii) T
D   (i) F   (ii) F

**6**   Decide whether each of these statements is TRUE (T) or FALSE (F).

A mechanically hard material will:
(i) Easily be dented.
(ii) Scratch easily.

A   (i) T   (ii) T
B   (i) T   (ii) F
C   (i) F   (ii) T
D   (i) F   (ii) F

Questions 7 and 8 relate to the Wheatstone bridge shown in Figure 3.41.

**7**   Decide whether each of these statements is TRUE (T) or FALSE (F).

When the bridge is balanced:
(i)   The current through AC *must* be the same as the current through AD.
(ii) The current through AC *must* be the same as the current through CB.

A   (i) T   (ii) T
B   (i) T   (ii) F
C   (i) F   (ii) T
D   (i) F   (ii) F

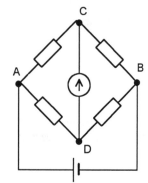

Figure 3.41   *Wheatstone bridge*

**8**   Decide whether each of these statements is TRUE (T) or FALSE (F).

When the bridge is balanced:
(i)   The potential difference between A and C *must* be the same as the potential difference between A and D.
(ii) The potential difference between C and B *must* be the same as the potential difference between D and B.

A   (i) T   (ii) T
B   (i) T   (ii) F
C   (i) F   (ii) T
D   (i) F   (ii) F

**9** In an experiment to determine the specific heat capacity of a liquid, an electrical heater is immersed in the liquid and the rise in temperature measured when the heater is switched on for a measured amount of time.

Decide whether each of these statements is TRUE (T) or FALSE (F).

When the experiment was first tried, the rise in temperature was found to be small. To improve the accuracy of the measurement it was considered that the rise in temperature should be increased. The temperature rise is increased by:
(i) Increasing the mass of the liquid used.
(ii) Increasing the voltage applied to the heater.

A  (i) T  (ii) T
B  (i) T  (ii) F
C  (i) F  (ii) T
D  (i) F  (ii) F

**10** Decide whether each of these statements is TRUE (T) or FALSE (F).

In an experiment to measure the refractive index of a liquid by means of the air cell method, the angle at which the cell has to be set for no transmission of light is:
(i) Increased if the thickness of the air layer in the cell is increased.
(ii) Increased if the thicknesses of the glass sheets sandwiching the air layer are increased.

A  (i) T  (ii) T
B  (i) T  (ii) F
C  (i) F  (ii) T
D  (i) F  (ii) F

**11** An instrument has a scale with graduations at intervals of 0.1 units. What is the worst possible reading error?

**12** Repeated measurements of the forces necessary to break a tensile test specimen gave: 802, 799, 800, 798, 801 kN. Determine (a) the average force, and (b) the standard error of the mean.

**13** Repeated measurements of the resistance of a resistor gave: 51.1, 51.3, 51.2, 51.3, 51.7, 51.0, 51.5, 51.3, 51.2, 51.4 $\Omega$. Determine (a) the average resistance, and (b) the standard error of the mean.

**14** Repeated measurements of the voltage necessary to cause the break-down of a dielectric gave: 38.9, 39.3, 38.6, 38.8, 38.8, 39.0, 38.7, 39.4, 39.7, 38.4, 39.0, 39.1, 39.1, 39.2 kV. Determine (a) the average breakdown voltage, and (b) the standard error of the mean.

**15** The total resistance of two resistors in series is the sum of their resistances. Determine the worst possible error in the total resistance if the resistors are 50 $\Omega$ with 10% accuracy and 100 $\Omega$ with 5% accuracy.

**16** When two resistors are connected in parallel, the total resistance is given by $R = R_1 R_2/(R_1 + R_2)$. What will be the worst possible error in the total resistance if the resistors are 50 $\Omega$ with a 10% accuracy and 100 $\Omega$ with a 5% accuracy?

**17** The volume of a cube with sides of $L$ is $L^3$. If the length is measured as $121 \pm 2$ mm, determine the worst possible error in the volume.

**18** The density of a solid is its mass divided by its volume. If the mass is measured as $42.5 \pm 0.5$ g and the volume as $54 \pm 1$ cm$^3$, what will be the worst possible error in the calculated density?

**19** The following results were obtained from a tensile test of an aluminium alloy. The test piece had a diameter of 11.28 mm and a gauge length of 56 mm. Plot the stress–strain graph and determine the tensile modulus.

| Load (kN) | 0 | 2.5 | 5.0 | 7.5 | 10.0 | 12.5 | 15.0 | 17.5 |
|---|---|---|---|---|---|---|---|---|
| Ext. (mm) | 0 | 1.8 | 4.0 | 6.2 | 8.4 | 10.0 | 12.5 | 14.6 |

| Load (kN) | 20.0 | 22.5 | 25.0 | 27.5 | 30.0 | 32.5 | 35.0 |
|---|---|---|---|---|---|---|---|
| Ext. (mm) | 16.3 | 19.0 | 21.2 | 23.5 | 25.7 | 28.1 | 31.5 |

| Load (kN) | 37.5 | 38.5 | 39.0 | 39.0 | (broke) |
|---|---|---|---|---|---|
| Ext. (mm) | 35.0 | 40.0 | 61.0 | 86 | |

**20** The following results were obtained from a tensile test of a polymer. The test piece had a width of 20 mm, a thickness of 3 mm and a gauge length of 80 mm. Plot the stress–strain graph and determine the tensile strength.

| Load (N) | 0 | 100 | 200 | 300 | 400 | 500 | 600 | 650 | 630 |
|---|---|---|---|---|---|---|---|---|---|
| Ext. (mm) | 0 | 0.08 | 0.17 | 0.35 | 0.59 | 0.88 | 1.33 | 2.00 | 2.40 |

**21** The following results were obtained from a tensile test of a steel. The test piece had a diameter of 10 mm and a gauge length of 50 mm. Plot the stress–strain graph and determine (a) the tensile strength, (b) the yield stress, (c) the tensile modulus.

| Load (kN) | 0 | 5 | 10 | 15 | 20 | 25 | 30 | 32.5 | 35.8 |
|---|---|---|---|---|---|---|---|---|---|
| Ext. (mm) | 0 | 0.016 | 0.033 | 0.049 | 0.065 | 0.081 | 0.097 | 0.106 | 0.250 |

**22** The following data was obtained from a tensile test on a stainless steel test piece. Determine (a) the limit of proportionality stress, (b) the tensile modulus.

| Stress (MPa) | 0 | 90 | 170 | 255 | 345 | 495 | 605 |
|---|---|---|---|---|---|---|---|
| Strain ($\times 10^{-4}$) | 0 | 5 | 10 | 15 | 20 | 30 | 40 |

| Stress (MPa) | 700 | 760 | 805 | 845 | 880 | 895 |
|---|---|---|---|---|---|---|
| Strain ($\times 10^{-4}$) | 50 | 60 | 70 | 80 | 90 | 100 |

**23** The following are Izod impact energies at different temperatures for samples of annealed cartridge brass (70% Cu–30% Zn). What can be deduced from the results?

| Temperature (°C) | +27 | −78 | −197 |
|---|---|---|---|
| Impact energy (J) | 88 | 92 | 108 |

**24** The following are Charpy V-notch impact energies for annealed titanium at different temperatures. What can be deduced from the results?

| Temperature (°C) | +27 | −78 | −197 |
|---|---|---|---|
| Impact energy (J) | 24 | 19 | 15 |

**25** The following are Charpy impact strengths for nylon 6.6 at different temperatures. What can be deduced from the results?

| Temperature (°C) | −23 | −33 | −43 | −63 |
|---|---|---|---|---|
| Impact strength (kJ/m$^2$) | 24 | 13 | 11 | 8 |

**26** With the Vickers hardness test, a 30 kg load gave for a sample of steel an indentation with diagonals having mean lengths of 0.530 mm. What is the hardness?

**27** With the Vickers hardness test, a 30 kg load gave for a sample of steel an indention with diagonals having mean lengths of 0.450 mm. What is the hardness?

**28** With the Vickers hardness test, a 10 kg load gave for a sample of brass an indentation with diagonals having means lengths of 0.510 mm. What is the hardness?

**29** With the Brinell hardness test, a 10 mm diameter ball and 3000 kg load gave an indentation with a diameter of 4.10 mm. What is the hardness?

**30** With the Brinell hardness test, a sample of cold worked copper with a 1 mm diameter ball and 20 kg load gave an indentation of diameter 0.630 mm. What is the hardness?

**31** In an experiment to determine the specific heat capacity of a liquid, 140 g of it were contained in a calorimeter of mass 80 g. For 130 s a voltage of 12.0 V was applied to an electrical heater immersed in the stirred liquid, the current being 3.0 A. This raised the temperature of the liquid from 5°C below room temperature to 5°C above it. If the calorimeter has a specific heat capacity of 400 J kg$^{-1}$ K$^{-1}$, what is the specific heat of the liquid?

**32** In an experiment to determine the specific heat capacity of aluminium, an insulated block of mass 500 g was heated by an electrical heater for 100 s. The voltage across the heater was 12.0 V and the current through it 2.0 A. The temperature of the block was raised by 5.5°C. What is the specific heat of the aluminium?

**33** With the air cell method of determining the refractive index of a liquid, the cut-off of transmission is found to occur when the cell is at an angle of 46°. What is the refractive index of the liquid?

# 4 The structure of solids

## 4.1 Structure and properties

Why are metals good conductors of electricity? Why are polymers weaker and less stiff than metals? Why are ceramics so stiff? How can we make a metal more ductile? In order to answer these and other questions, and gain an understanding of the properties of materials, we need to consider materials at the atomic level. All solids are made up of particles consisting of single atoms, ions or molecules. Thus, consideration of the forces between the particles and the ways in which the particles are packed in solids leads to an understanding of mechanical properties and so how materials behave when stretched or compressed. Consideration of the charge available for electrical conduction through solids gives an understanding of electrical conductivity.

This chapter is about the basic form of the structure of metals, polymers and ceramics. The following chapter indicates how the properties of such materials can be explained in terms of structure and modified by structural changes.

## 4.2 Forces between atoms and molecules

Consider what forces there must be between atoms and molecules in a solid. A solid has a fixed shape and does not spontaneously change its shape, so there must be no resultant force acting on the particles constituting the solid. Only if there is no resultant force will the particles not move. The particles all stick together in the solid form, so there must be forces of attraction that pull the particles together. But if we squeeze a solid, it resists being squashed. There must therefore be forces of repulsion that oppose the particles being forced further together.

We can think of the situation in a solid as being like that of an array of particles with each particle linked to its neighbours by springs (Figure 4.1). The particles are held together by attractive forces exerted on each of them by the spring. If you try to pull the particles further apart, then the spring exerts attractive forces on the particles to pull them back to their normal separation. However, if we apply forces on the particles to push them closer together, the spring exerts repulsive forces. The more we try to push the particles together, the greater the repulsive force. We can thus think of their being attractive forces and repulsive forces which vary with separation between particles. At the normal separation of the particles, they neither move further in or out and so the attractive forces must be just cancelled by the repulsive forces. At greater separations, the attractive force predominates and at closer separations the repulsive force predominates.

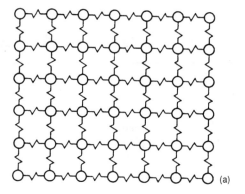

Figure 4.1 *A simple model of a solid*

Figure 4.2 shows how we might expect these forces to vary with the separation distance between particles in a solid. The force of attraction decreases as the separation increases. The force of repulsion also decreases as the separation increases, but at a faster rate. At the normal separation of the particles in a solid, the repulsive force and attractive force will cancel so there is no resultant force acting on the particles. They are in equilibrium. The graph shows how the resultant force, i.e. the sum of the attractive and repulsive forces, varies with separation. For separations greater than the normal separation, there is a resultant attractive force which acts to pull the particles back to the normal separation. With separations less than the normal separation, there is a resultant repulsive force which pushes the particles back out to the normal separation.

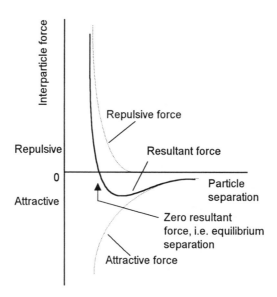

Figure 4.2  *Forces between particles*

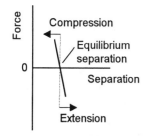

Figure 4.3  *Hooke's law*

If we consider the resultant force on the particles for separations close to their equilibrium separation (Figure 4.3), then the force is reasonably proportional to the amount by which we extend or compress the particles together from their equilibrium position. In other words, the extension or compression is proportional to the applied force and we have Hooke's law.

Consider two particles exerting attractive forces on each other. If we start to increase the separation of the particles then we have to do work. Think of it being like stretching a spring between them. The further we move one particle from the other, the more work we have to do. Thus when one of the particles has been moved to an infinite distance from the other, then the maximum amount of work would have been done. When a particle on the end of a tethered spring is pulled so that the spring stretches, then work is done and the particle gains potential energy. If you let go, the particle on the spring springs back, loosing the potential energy given to it by the stretching. Thus, when two particles, between which there are attractive forces, are pulled apart, they gain potential energy and when one has been moved to an infinite distance from the other, then we would expect the potential energy to be a maximum. However, it is convenient to regard the potential energy of each particle to be zero when they are an infinite distance apart. This is because at such a separation they have no influence on each other. As a consequence of this, when there is a resultant force of attraction between two particles their potential energy is a negative quantity. Only with it negative can it be said to increase when the separation is increased to infinity.

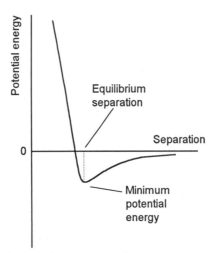

Figure 4.4 *Interparticle energy*

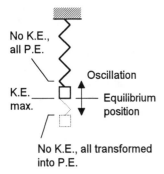

Figure 4.5 *Oscillation*

Figure 4.4 shows how the potential energy between atoms or molecules in a solid varies with the separation between them. At the equilibrium separation, i.e. the separation when there is no resultant force, the potential energy is a minimum. The deeper this 'potential energy well', the greater the energy that has to be supplied to separate the particles completely in the solid. Thus the greater the *binding energy*.

When a material is heated, its particles gain energy. In a gas where the molecules are so far apart that usually we can ignore intermolecular forces for all but the small amounts of time when they collide, the result of heat energy being supplied is to increase the kinetic energy of the molecules so that they move around faster. In a solid the interparticle forces are such that the particles are not free to move about as molecules can in a gas. What we can, however, consider to happen is that the influx of heat energy causes the particles to oscillate about their equilibrium positions. With an oscillating mass on the end of a tethered spring (Figure 4.5), when the mass is at the end of its travel and the maximum distance away from its equilibrium position and momentarily has zero velocity, all its energy is in the form of potential energy. When the oscillating mass is passing through its equilibrium position its velocity is a maximum and potential energy has been transformed into kinetic energy. In the case of a solid, when it is at the absolute zero of temperature, it has no kinetic energy and thus the particles are in their equilibrium positions with the minimum potential energy value. At temperatures above the absolute zero, there is an influx of heat energy to the solid and so the particles will have kinetic energy which results in them oscillating with this energy being transformed into potential energy and then back into kinetic energy, and so on repeatedly. The particle thus oscillates between the points A and B on the potential energy graph (Figure 4.6). If there is a bigger input of heat energy, i.e. the solid is raised to a higher temperature, then the particle oscillates between the points C and D, higher up the potential energy graph than points A and B. The midpoint of these oscillations moves to greater separations as the temperature increases. Thus as the temperature of the solid increases, the solid expands.

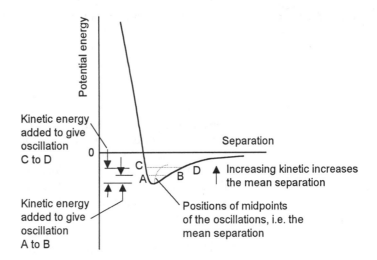

Figure 4.6 *Expansion*

**4.3 Bonds**  Atoms are held together in solids and molecules by electrical forces. An electric force of attraction occurs between positively and negatively charged particles, a force of repulsion between two positively or two negatively charged particles. There are various ways in which such electric forces can produce the forces between particles in solids, i.e. the *interparticle bonds*. These bonds can be classified in two groups:

1  *Primary bonds.* These are *metallic, ionic* and *covalent* bonds, and all are relatively strong bonds.
2  *Secondary bonds.* These are *van der Waals* and *hydrogen* bonds. Both are relatively weak.

Metals and ceramics are held together by primary bonds, mainly the metallic bond for metals and the ionic and covalent bonds for ceramics. These strong bonds are responsible for the high strength and stiffness of these materials. Secondary bonds provide the bonds responsible for holding together solid polymer materials. Their relatively weak nature means that such materials are less strong and less stiff.

**4.3.1 The metallic bond**

The term *metal* is used for elements, such as copper, which have atoms which so readily loose electrons that in the solid state at room temperature there are many free electrons. Thus in the solid state, copper consists of an

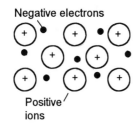

Negative electrons

Positive
ions

Figure 4.7 *Metallic bonding*

array of atoms each of which has lost one electron (Figure 4.7). This leaves each copper atom as having a net positive charge and it is termed a positive ion. The electrons that have been lost do not combine with any one ion but remain as a cloud of negative charge floating between the ions. The result is rather like a glue in that the cloud of electrons holds the positive ions together, positive ions being attracted to electrons which in turn attract other positive ions. This is what is termed the *metallic bond*. It is the dominant, though not the only, bond in metals.

In a metal we will have the attractive forces between positive ions and the electrons and repulsive forces between the positive ions. At the normal separation of the metal ions, the forces of attraction are just balanced by the forces of repulsion. If we try to compress a metal and so move the positive ions closer together, the repulsive force predominates. If we try to stretch a metal, the attractive force predominates. Since the bonds formed between the positive ions can be formed in any direction without any restrictions, a simple model we can use to describe the structure of metals is to think of the ions in a metal being like spheres, a sphere being a shape which imposes no directionality rules on how they are packed together.

The free electrons explain why metals are good conductors of electricity, since they have free charge carriers which are easily moved through the solid by the application of a voltage. Insulators have no free electrons and the atoms in the solid are bonded together in a different way.

### 4.3.2 The ionic bond

An individual atom is electrically neutral, having as much positive charge in the nucleus as negative charge in its electrons. However, if an atom loses an electron, it must then have a net positive charge. It is then referred to as an *ion*, in this case a positive ion. If an atom gains an extra electron, it ends up with a net negative charge, becoming a negative ion. In an ionically bonded solid, such as a block of common salt (sodium chloride), an atom of one element looses an electron to an atom of the other element. The sodium looses and electron to the chlorine. The result is that we have a positive ion and a negative ion. Unlike-charged particles attract each other. Thus there is a force of attraction and this is what is referred to as the *ionic bond* (Figure 4.8).

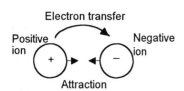

Electron transfer

Positive
ion

Negative
ion

Attraction

Figure 4.8 *Ionic bonding*

The solid sodium chloride exists as not just a pair of ions but a vast structure of sodium and chlorine ions (Figure 4.9). In such an array, unlike-charged ions attract each other and like-charged ions repel each other. Thus sodium ions repel sodium ions, chlorine ions repel chlorine ions, but sodium and chlorine ions attract each other. In this way we can explain both the attractive and repulsive forces shown in Figure 4.2. A stable structure exists because the arrangement of the sodium and chlorine ions are such that the attractive forces are just balanced by the repulsive forces.

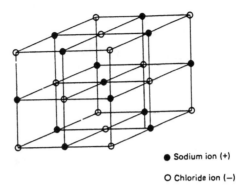

● Sodium ion (+)

O Chloride ion (−)

Figure 4.9 *Sodium chloride*

The bonds between ions can be formed in any direction, there being no requirements for ions to be lined up in any particular way. There is thus no directionality of bonding.

Because the electrons are all tied up in bonds, there are no free electrons to act as the charge carriers for electrical currents. Thus materials with ionic bonds are electrical insulators. Because ionic bonds are strong bonds, such materials tend to have high melting points, a lot of thermal energy being needed to break the points.

### 4.3.3 The covalent bond

Covalent bonding can be considered to be where neighbouring atoms share electrons. Figure 4.9 illustrates this in a simplistic manner. We can think of a shared electron as being in orbit about the pair of atoms, rather than just one of the atoms. As a result we can think of the result being a greater chance of the shared electrons being between the atoms and so acting as a 'glue' holding together the atoms. We have a positive charge attracted towards the shared electrons which in turn attracts the other positive charge.

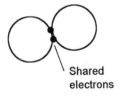

Shared
electrons

Figure 4.9 *Covalent
bonding*

An atom can share electrons with more than one other atom. For example, a carbon atom may share electrons with four hydrogen atoms, the result being a methane molecule. Each hydrogen atom shares one of the carbon electrons and the carbon atom obtains a share in one electron from each hydrogen atom. There are thus four covalent bonds. Figure 4.10(a) gives a two-dimensional picture of this methane molecule. However, the covalent bond is highly directional and so a more realistic representation of the methane molecule is given by the three-dimensional figure shown in Figure 4.10(b).

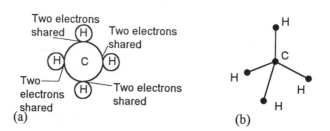

Figure 4.10 *Methane*

Covalent bonding is the dominant type of bond in silicate ceramics. It is a strong bond and thus such materials have high melting points, i.e. a lot of thermal energy is needed to break the bonds. Because the electrons are all tied up in bonds, there are no free electrons to act as the charge carriers for electrical currents. Thus materials with covalent bonds are electrical insulators.

### 4.3.4 van der Waals bonds

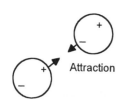

Figure 4.11 *Van der Waals bonding*

We can think of an atom as having a positively charged nucleus around which negatively charged electrons orbit. With an atom having no net charge, the charge on the nucleus is just balanced by the charge carried by the electrons. We might thus expect that, when we have two atoms with no net charge, there will be no bonding forces between them. However, the distribution of the negative charge about an atom can vary with time. Thus, at some instant of time, we might have an atom with more negative charge on one side than the other. Some molecules might have a permanent polarisation of charge in this way with some parts with a net negative charge and some with a positive charge. We can thus have electrostatic forces of attraction between such polarised atoms or molecules. The situation becomes something like that shown in Figure 4.11. This is the force responsible for what is termed *van der Waals bonding*.

This is the force responsible for the bonding between the polymer molecules in a polymer. It is a much weaker force than the covalent bonds that are usual among the atoms in the polymer molecules.

### 4.3.5 Hydrogen bonding

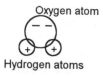

Figure 4.12 *The water dipole*

The water molecule consists of two oxygen atoms and a hydrogen atom. When water freezes to form a solid, we must have bonds formed to hold the material together. The covalent bonding in the water molecule results in the shared electrons being closer to the oxygen atom than the hydrogen atoms. This results in the two hydrogen ends of the molecule being slightly positive relative to the oxygen atom. Each water molecule is said to be a *dipole* (Figure 4.12). Hydrogen bonding occurs as a result of electrostatic attraction between these molecular dipoles. The cell walls of plants, and hence wood, are made of a naturally occurring polymer called cellulose. The cellulose molecules are bound to each other by hydrogen bonds.

## 4.4 Crystals and amorphous structures

Consider the stacking together of spheres in an orderly manner. Since spheres can be stacked in any way, a sphere can be considered to be a model for an atom, ion or molecule in a solid when there is no directionality to the bonding forces.

One of the simplest arrangement of spheres is that of the *simple cubic structure*. Figure 4.13 shows the structure obtained by stacking four spheres with the centres of each spheres at the corners of a cube. The surfaces of each sphere touch the surfaces of each of its neighbours in such a way that the length of the side of the cube is equal to the diameter of the spheres. The dotted line in the figure encloses what is termed the *unit cell*, this being the smallest arrangement of particles that when regularly repeated forms the crystal. The resulting solid would consist of a completely orderly array of spheres, i.e. particles, and we would expect the surfaces of such a solid to be smooth and flat with the angles between adjoining faces always 90°. Such a solid would when broken up always have the appearance of stacked cubes. This is a description of a *cubic crystal*.

The simple cubic crystal shape is arrived at by stacking spheres in one particular way. However, it is not the way that spheres can be most closely packed. By stacking spheres in a closer manner, as in Figure 4.14, other crystal shapes can be produced. With the *body-centred cubic* unit cell (Figure 4.14(a)), the arrangement is slightly more complex than the simple cubic unit cell in having an extra sphere in the centre of the cell. With the *face-centred cubic* unit cell (Figure 4.14(c)), there is, when compared with the simple cubic unit cell, a sphere at the centre of each face of the cube. With the *hexagonal close-packed* unit cell (Figure 4.14(b)), the spheres are packed in a close array which gives a hexagonal form of structure. These three close-packed structures represent the structures occurring with solid metals.

An important point to notice with all these structures is that there are spaces between the spheres in the crystal structures. The size of these spaces depends on the type of structure. Within these spaces it is possible to fit other atoms, provided they are small enough, without too much strain on the crystalline structure. In some cases, with some strain, atoms can be forced into spaces which are really too small for them. This is discussed in the next chapter in connection with alloys. A crystal thus consists of a large number of particles arranged in a regular repetitive array. It is this regularity that is characteristic of crystalline material.

With crystalline structures the atoms, ions or molecules are arranged in a regular, orderly manner. A solid having no such order in the arrangement of its constituent particles is said to be *amorphous*.

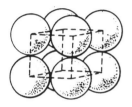

Figure 4.13 *A simple cubic structure*

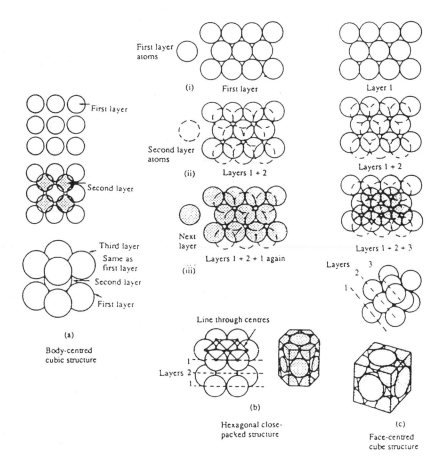

Figure 4.14 *Stacking spheres*

### 4.4.1 Metals as crystalline materials

Metals are crystalline substances. This may seem a strange statement in that metals do not generally seem to look like crystals, with their geometrically regular shapes. However, if we consider a metal in solidifying from the liquid as not growing as a single crystal but having crystals starting to grow at a large number of points within the liquid, the result is a mass of crystals. Each crystal in growing is prevented from reaching geometrically regular shapes by neighbouring crystals restricting its growth.

The term *grain* is used to describe the crystals within the metal. A grain is merely a crystal without its geometrical shape and flat surfaces because its growth was impeded by contact with other crystals. Within a grain, the arrangement of particles is just as regular and repetitive as within a crystal with smooth faces. A simple model of a metal with grains is given if a raft of bubbles is produced on the surface of a liquid. Figure 4.15 shows how such bubbles can be produced and an example of the type of bubble raft that might be produced. The bubbles pack together in an orderly and repetitive manner but if 'growth' is started at a number of centres, then 'grains' are produced. At the boundaries between the 'grains', the regular pattern breaks down as the pattern changes from the orderly pattern of one 'grain' to that of the next 'grain'.

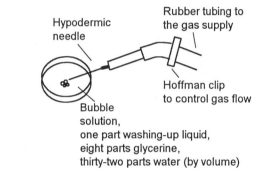

(a)

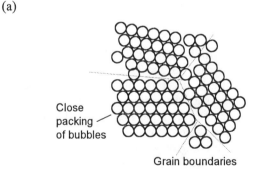

(b)

Figure 4.15  *(a) Simple arrangement for producing bubbles, (b) grains in a bubble raft*

The grains in the surface of a metal are not generally visible, though an exception occurs with the very large grains which are readily visible in the surface of galvanised steel objects. Grains can, however, be made visible by careful etching of the polished surface of the metal with a suitable chemical. The chemical preferentially attacks the grain boundaries. For example, in the case of copper and its alloys, concentrated nitric acid can be used. In the case or carbon and alloy steels of medium carbon content, a etchant called nital can be used. Nital is a mixture of nitric acid and alcohol, typically 5 ml of acid to 95 ml of alcohol. However, proper safety precautions in the handling and disposing of the chemicals are vital since they are highly corrosive and many of them are potentially lethal.

Examples of metals that, in the pure state, adopt the body-centred cubic unit cell structure are iron, chromium and molybdenum, face-centred cubic unit cell structure aluminium, copper, lead and nickel, with the hexagonal close-packed unit cell being given by magnesium and zinc.

You can study the growth of grains in a simple experiment with materials such as phenyl salicylate (phenyl-2-hydroxy benzoate). This material crystallises in a similar way to metals but melts at a temperature (43°C) just above room temperature and is transparent. Warm, in an oven, two pieces of glass, e.g. microscope slides, to about 50°C for about 15 minutes. Warm the phenyl salicylate in a test-tube until is just melts and then pour a few drops of the liquid on to the warm glass. Cover the liquid with the second piece of glass so that a thin layer of liquid is between the sheets of glass. Observe the growth of the crystals as the liquid cools.

## 4.5 The structure of polymers

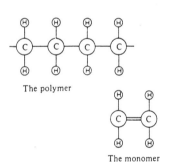

The polymer

The monomer

Figure 4.16 *Polyethylene*

The plastic washing-up bowl, the plastic measuring rule, the plastic cup are all examples of materials that have polymer molecules as their basis. A polymer molecule in a plastic may have thousands of atoms all joined together in a long chain. The backbones of these long molecules are chains of atoms linked together by covalent bonds. The chain backbone is usually predominantly carbon atoms and what we have are long molecules formed by the repetition of basic structural units formed by groups of atoms, i.e. a small bit of a chain. The term *polymer* is used to indicate that a compound consists of many repeated structural units. The prefix 'poly' means many. Each structural unit in the compound is called a *monomer*. For many plastics the monomer can be deduced by deleting the prefix 'poly' from its name. Thus the plastic called polyethylene is a polymer which has as its monomer ethylene. Figure 4.16 shows the monomer and part of the resulting polymer chain.

Figure 4.17 shows the basic forms that can be adopted by the molecular chains. These forms can be described as *linear chains*, *branched chains* and *cross-linked chains*. A solid polymer may thus consist of linear chains arranged in some way with the bonding between them by van der Waals bonding. The linear chains have no side branches or ionic or covalent cross-links with other chains and can thus move readily past each other, breaking and remaking van der Waals bonds. If, however, the chains have side branches, there is a reduction in the ease with which chains can move past each other, i.e. the van der Waals bonds be broken, and so the material is more rigid. If there are ionic or covalent cross-links, a much more rigid material is produced in that the chains cannot slide past each other at all and the solid may be considered to be almost just one large cross-linked chain.

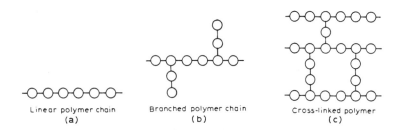

Linear polymer chain
(a)

Branched polymer chain
(b)

Cross-linked polymer
(c)

Figure 4.17 *(a) Linear polymer chain, (b) branched polymer chain, (c) cross-linked polymer chain*

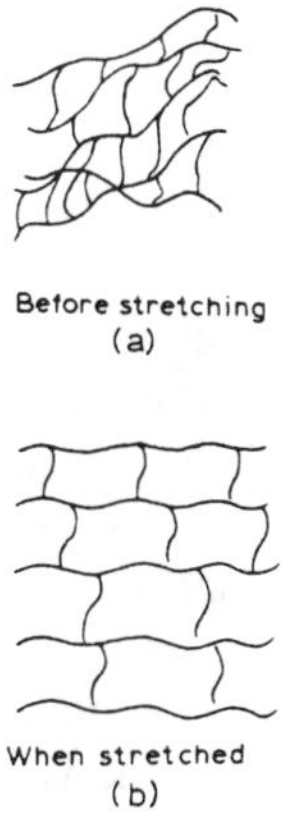

Polymers can be classified as *thermoplastics, thermosets* or *elastomers*. A simple method by which thermoplastics and thermosets can be distinguished is when heat is applied. With a thermoplastic the material softens with removal of the heat resulting in hardening. With a thermoset, heat causes the material to char and decompose, with no softening. An elastomer is a polymer that by its structure allows considerable extensions which are reversible. Thermoplastics have linear chains or branched chains for their structure. Thermosets have a cross-linked structure. Elastomers are chains with some degree of cross-linking.

The atoms in a thermoset form a three-dimensional structure of chains with frequent cross-links between chains (Figure 4.17(c)). The bonds linking the chains are strong and not easily broken. Thus the chains cannot slide over one another. As a consequence, thermosetting polymers are stronger and stiffer than thermoplastics. Thermoplastics offer the possibility of being heated and then pressed into the required shapes. Thermosets cannot be so manipulated. The processes by which thermosetting polymers can be shaped are limited to those where the product is formed by the chemicals being mixed together in a mould so that the cross-linked chains are produced while the material is in the mould. The result is a polymer shaped to the form dictated by the mould. No further processes, other than possibly some machining, are likely to occur.

Elastomers are polymers that can show very large, reversible strains when subject to stress. The behaviour of the material is perfectly elastic up to considerable strains, e.g. you can stretch a rubber band up to more than five times its unstrained length and it is still elastic. Elastomers have a structure consisting of tangled polymer chains which are held together by occasional cross-linked bonds. The difference between thermosets and elastomers is that with thermosets, there are frequent cross-linking bonds between chains while with elastomers there only occasional bonds. A simple model for the elastomer structure might be a piece of very open netting (Figure 4.18). In the unstretched state the netting is in a loose pile. In the elastomer, there will be some weak temporary bonds, van der Waals bonds, between chains in close proximity to each other, these being responsible for holding the

Before stretching
(a)

When stretched
(b)

Figure 4.18 *An elastomer: (a) unstretched, (b) stretched*

tangled chains together. When the material begins to be stretched the netting just begins to untangle itself and large strains can be produced. The van der Waals bonds between the chains will cause the elastomer to spring back to its original tangled state when the stretching forces are removed. It is not until quite large strains are applied, when the netting has become fully untangled and the structure is orderly, that the bonds between atoms in the materials begin to be significantly stretched. At this point the material becomes more stiff, i.e. stress–strain graph starts to become more steep, and much larger stresses are needed to give further extensions.

### 4.5.1 Crystalline and amorphous materials

A crystalline structure is one in which there is an orderly arrangement of particles; a structure in which the arrangement is completely random is said to be amorphous. Many polymers are amorphous with the polymer chains being completely randomly arranged in the material. Figure 4.19 illustrates this, the chains being shown as lines, individual atoms not being indicated. Linear polymer molecules can, however, assume an arrangement which is, at least partially, orderly. Figure 4.20 shows the type of arrangement of chains that can occur, the linear chains folding backwards and forwards on themselves. The arrangement is said to be *crystalline*.

The tendency of a polymer to crystallise is determined by the form of the polymer chains. Linear polymers can crystallise to quite an extent, complete crystallisation is not, however, obtained in that there is invariably some regions of disorder. For example, linear polyethylene chains (Figure 4.21(a)) can have some 95% of the material crystalline. Polymers with side branches show less tendency to crystallise since the branches get in the way of the orderly arrangement. For example, polyethylene in a branched form (Figure 4.21(b)) might show crystallinity in about 50% of the material. The greater the crystallinity of a polymer, the closer the polymer chains can be packed and so the greater the density of the solid polymer. The term *high density polyethylene* is used for the polyethylene with crystallinity of about 95%, *low density polyethylene* for that with crystallinity of about 50%. The high density polyethylene has a density of about 950 kg/m³, the low density about 920 kg/m³. The closer packing of the chains in the high density polyethylene means that there can be more inter-chain van der Waals bonds, hence a higher melting point (138°C) than the low density form (115°C).

When an amorphous polymer is heated, it shows no definite melting temperature but progressively becomes less rigid. This is because the arrangement of the chains in the solid is disorderly, just like in a liquid, and so there is no structural change occurring at melting. With a crystalline polymer, there is an abrupt change in structure at a particular temperature when the crystalline structure changes to a disorderly structure. If the density of the polymer were being monitored, there would be an abrupt change in density when this occurs as a result of a change in the way the chains are packed together. This temperature is termed the *melting point*.

Figure 4.19 *A linear amorphous polymer*

Figure 4.20 *Folded linear polymer chains*

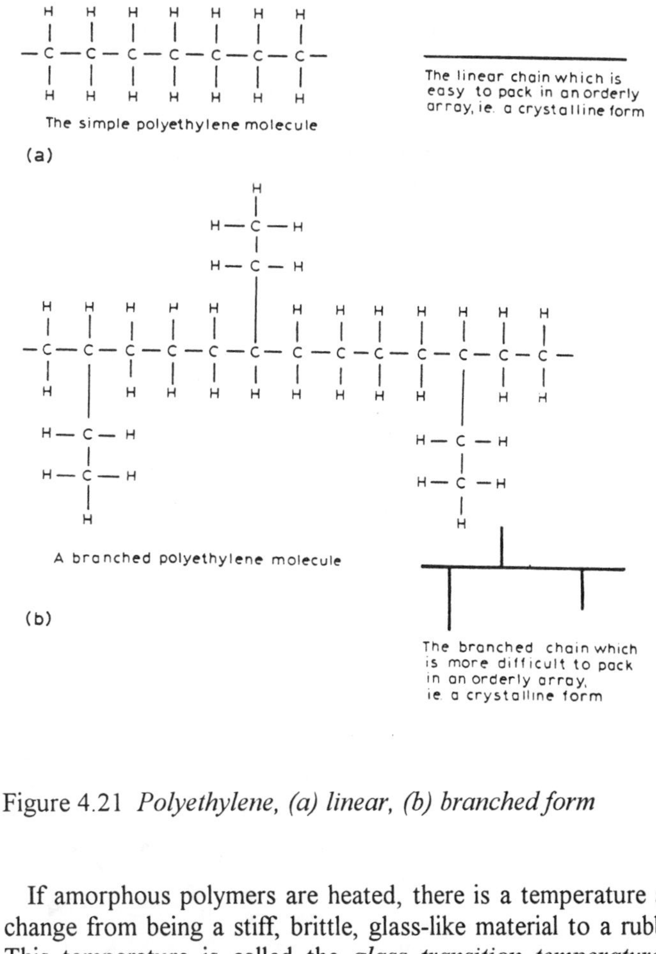

H   H   H   H   H   H   H
|   |   |   |   |   |   |
— C — C — C — C — C — C — C —
|   |   |   |   |   |   |
H   H   H   H   H   H   H

The simple polyethylene molecule

The linear chain which is
easy to pack in an orderly
array, ie. a crystalline form

(a)

A branched polyethylene molecule

The branched chain which
is more difficult to pack
in an orderly array,
ie. a crystalline form

(b)

Figure 4.21 *Polyethylene, (a) linear, (b) branched form*

If amorphous polymers are heated, there is a temperature at which they change from being a stiff, brittle, glass-like material to a rubbery material. This temperature is called the *glass transition temperature*. Below this temperature, segments within the molecular chains are unable to move and the material is stiff with a high Young's modulus and generally rather brittle. Above this temperature, there is sufficient thermal energy for some motion of segments of the chains to occur. The material then becomes less stiff with a lower Young's modulus and more like an elastomer and rubbery (Figure 4.22). Perspex is an example of a glassy polymer at room temperature. It has a glass transition temperature of 120°C. Thus when it is heated above this temperature it can be easily bent and twisted. Below that temperature it is much stiffer and fairly brittle. This property can be used to enable such polymers to be moulded into shapes required for products.

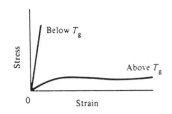

Figure 4.22 *Stress–strain graphs for a polymer below and above its glass transition temperature*

## 4.6 The structure of ceramics

Ceramics are combinations of one or metals, or semimetallic elements such as silicon, with a non-metallic element, usually oxygen. The bonds between the atoms are ionic or covalent. Because such bonds are strong, ceramics have high melting points. The combination of oxygen with metal atoms by ionic bonds gives strong bonds because each oxygen atom takes two electrons from its neighbouring metal atoms. The result is a strong electric force of attraction between the resulting ions.

Silica, i.e. $SiO_2$, forms the basis of a large range of ceramics. A silicon atom forms covalent bonds with four oxygen atoms (Figure 4.23(a)) to give a tetrahedron-shaped structure (Figure 4.23(b)). This structure leaves the oxygen atoms with 'spare' bonds with which they can link up with other silicon–oxygen tetrahedra or metal ions.

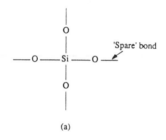

(a)

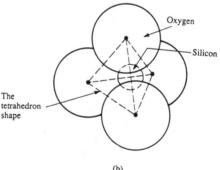

(b)

Figure 4.23  *(a) The silicon and oyygen bonds, (b) the tetrahedron-shaped silicon–oxygen structure*

When the tetrahedra link with other tetrahedra, because they only share corners, a range of structures can be assembled. With the tetrahedra are arranged in an orderly manner, a crystalline structure results (Figure 4.24), quartz is such a structure.

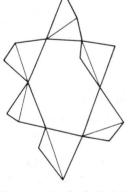

Figure 4.24  *Orderly linking of tetrahedra, this structure being repeated in a regular manner*

When silica is melted, a highly viscous liquid is produced. This viscosity arises because of the presence of strings of linked tetrahedra. These strings easily become tangled and prevent the liquid flowing easily. If silica in the molten state is cooled very slowly, crystalline structures are formed. However, if the molten silica is cooled more rapidly, it is unable to get all the tetrahedra into the orderly array required of a crystal and the resulting solid is a disorderly structure termed a *glass*. Thus silica glass is a disorderly arrangement of linked tetrahedra, i.e. an amorphous structure (Figure 4.25).

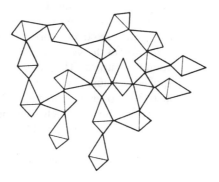

Figure 4.25 *Silica glass*

Silica is an important constituent of general usage glasses. Adding sodium oxide to molten silica reduces the viscosity of the molten silica by limiting the extent to which the tetrahedra link to each other. The sodium ions form ionic bonds with 'spare' tetrahedra bonds and prevent them linking to other tetrahedra. The result is a solid glass with less linked tetrahedra and hence a glass, soda-glass, which has a lower softening temperature. Other types of silica glass can be formed by using other oxides. For example, iron results in a green coloured glass, copper a blue-green colour, boron a glass with a low thermal expansion, lead a glass with a high refractive index.

## 4.6 Electrical properties of solids

In terms of their electrical conductivity, materials can be grouped into three categories, namely conductors, semiconductors and insulators. Conductors have electrical conductivities of the order of $10^6$ S/m, semiconductors about 1 S/m and insulators $10^{-10}$ S/m. Conductors are metals with insulators being polymers or ceramics. Semiconductors include silicon, germanium and compounds such as gallium arsenide. Silicon is the most widely used semiconductor.

In discussing electrical conduction in materials, a useful picture is of an atom as consisting of a nucleus surrounded by its electrons. The electrons are bound to the nucleus by electric forces of attraction. The force of

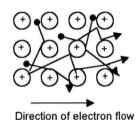

Direction of electron flow

Figure 4.26 *Electric current with a metal*

attraction is weaker the further an electron is from the nucleus. The electrons in the furthest orbit from the nucleus are called the valence electrons since they are the ones involved in the bonding of atoms together to form compounds.

*Metals* can be considered to have a structure of atoms with valence electrons which are so loosely attached that they drift off and can move freely between the atoms. Typically a metal will have about $10^{28}$ free electrons per cubic metre. Thus, when a potential difference is applied across a metal, there are large numbers of free electrons able to respond and give rise to a current. We can think of the electrons pursuing a zigzag path through the metal as they bounce back and forth between atoms (Figure 4.26). An increase in the temperature of a metal results in a decrease in the conductivity. This is because the temperature rise does not result in the release of any more electrons but causes the atoms to vibrate and scatter electrons more, so hindering their progress through the metal.

*Insulators*, however, have a structure in which all the electrons are tightly bound to atoms. Thus there is no current when a potential difference is applied because there are no free electrons able to move through the material. To give a current, sufficient energy needs to be supplied to break the strong bonds which exist between electrons and insulator atoms. The bonds are too strong to be easily broken and hence normally there is no current. A very large temperature increase would be necessary to shake such electrons from the atoms.

*Semiconductors* can be regarded as insulators at a temperature of absolute zero. However, the energy needed to remove an electron from an atom is not very high and at room temperature there has been sufficient energy supplied for some electrons to have broken free. Thus the application of a potential difference will result in a current. Increasing the temperature results in more electrons being shaken free and hence an increase in conductivity. At about room temperature, a typical semiconductor will have about $10^{16}$ free electrons per cubic metre and $10^{16}$ atoms per cubic metre with missing electrons.

Silicon is a covalently bonded solid with, at absolute zero, all the outer electrons of every atom involved in bonding with other atoms. Thermal shaking of the atoms results in some of the bonds breaking and freeing electrons. When a silicon atom looses an electron, we can consider there to be a hole in its valence electrons (Figure 4.27(a)). When electrons are made to move as a result of the application of a potential difference, i.e. an electric field, they can be thought of as hopping from valence site into a hole in a neighbouring atom, then to another hole, etc. Not only do electrons move through the material but so do the holes, the holes moving in the opposite direction to the electrons. We can think of the above behaviour in the way shown in Figure 4.27(b). One way of picturing this behaviour is in terms of a queue of people at, say, a bus-stop. When the first person gets on the bus, a hole appears in the queue between the first and second person. Then the second person moves into the hole, which now moves to between the second and third person. Thus as people move up the queue, the hole moves down the queue.

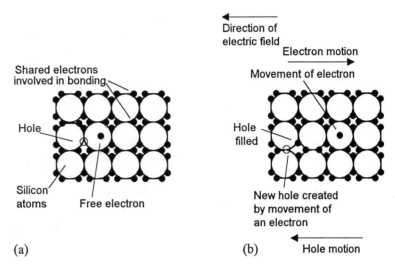

Figure 4.27 *(a) Holes and free electrons in silicon, (b) movement of holes and free electrons when an electric field is applied*

The conductivity of a semiconductor can be very markedly changed by impurities. For this reason the purity of semiconductors must be very carefully controlled. With the silicon used for the manufacture of semiconductor devices, the impurity level is routinely controlled to less than one atom in a thousand million silicon atoms. Foreign atoms can, however, be deliberately introduced in controlled amounts into a semiconductor in order to change its electrical properties. This is referred to as *doping*. Atoms such as phosphorus, arsenic or antimony when added to silicon add easily-released electrons and so make more electrons available for conduction. Such dopants are called *donors*. Semiconductors with more electrons available for conduction than holes are called an *n-type semiconductor*. Atoms such as boron, gallium, indium or aluminium add holes into which electrons can move. They are thus referred to as *acceptors*. Semiconductors with an excess of holes are called a *p-type semiconductor*.

### 4.6.1 Dielectrics

An electric field is produced in a material when a potential difference is applied across it. Charged particles in electric fields experience forces, so if an electric field is produced in a conductor, the free electrons in it move and a current occurs. Insulators, however, have no free electrons and thus when an electric field is produced there is no movement of free electrons. But positive and negative charges within the particles of the material will be acted on by forces and can become displaced slightly.

The term *dipole* is used for atoms or groups of atoms that effectively have a positive charge and a negative charge separated by a distance (Figure 4.28). These may be permanent dipoles because of an uneven

Figure 4.28 *A dipole*

Figure 4.29 *Permanent dipole due to uneven charge distribution*

distribution of charge in a molecule (Figure 4.29). The material is said to show *molecular polarisation*. When an electric field is applied to a material containing permanent dipoles, the dipoles become reasonably lined up with the field (just like compass needles line up with a magnetic field). Figure 4.30 illustrates this, the material being between the plates of a parallel plate capacitor and the electric field produced by a potential difference applied between the plates.

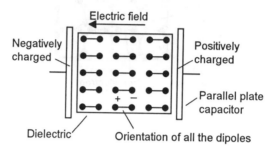

Figure 4.30 *Dipoles in an electric field*

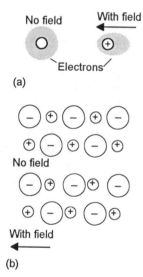

Figure 4.31 *Production of temporary dipoles*

With some materials, dipoles may also be temporally created when an electric field is applied. This may occur as a result of the electric field distorting the arrangement of the electrons in orbit about the nucleus (Figure 4.31(a)). This is termed *electronic polarisation*. With an ionic bonded material, such as sodium chloride, since such materials consist of an orderly array of positive and negative ions and the arrangement is completely symmetrical there is, in the absence of an electric field, no permanent dipole. However, when an electric field is applied, forces act on the ions and pull the positive ions slightly in one direction and the negative ions slightly in the opposite direction. This resulting distortion results in temporary dipoles (Figure 4.31(b)). This is known as *ionic polarisation*. In a dielectric such as polystyrene, the predominant mode of polarisation is electronic polarisation. With crystalline ceramics such as alumina, the predominant mode of polarisation is ionic polarisation.

When there is no dielectric but just a vacuum between the plates of a capacitor, when a potential difference is applied across the plates they become charged (Figure 4.32(a)). The amount of charge $Q_0$ on the plates is determined by the potential difference $V$ and is given by

$$Q_0 = C_0 V$$

where $C_0$ is the capacitance with a vacuum. The result of using a dielectric between the plates of the capacitor, whether the dipoles are permanent or temporary, is to give some alignment of the dipoles with the electric field and the charge on the plates is partially cancelled by the charge on the dipoles adjacent to the plates (Figure 4.32(b)). Thus there is a net smaller charge $Q$ on the plates for the same potential difference $V$. Since:

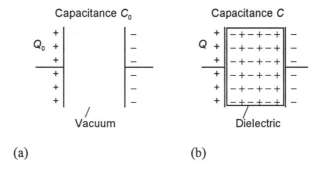

Figure 4.32 *Capacitor with (a) a vacuum, (b) a dielectric between the plates*

$$Q = CV$$

where $C$ is the capacitance of the plates with the dielectric present, then a lower $Q$ for the same $V$ must mean a higher value of $C$. The factor by which the capacitance is increased is the *relative permittivity* $\varepsilon_r$:

$$C = \varepsilon_r C_0$$

The relative permittivity is thus a measure of the polarisation, the greater the polarisation, the greater the relative permittivity.

If a dielectric material behaved perfectly, when the electric field was reversed, the dipoles would all instantly follow the change and become realigned. However, with real materials the particles in the dielectric are held together by bonds which resist reorientation. Thus the rotation of dipoles is slowed down. With low frequency alternating voltages applied across the plates of a capacitor, the dipoles have sufficient time to become reorientated. However, with higher frequency alternating voltages, this may not be the case. As a consequence, the relative permittivity depends on the frequency.

When a dielectric material is used with an alternating electric field, a fraction of the energy supplied is 'lost' each time the field is reversed. This loss is because the motion of reorientation is opposed by frictional affects in the dielectric. The energy dissipated depends on the material and the frequency, the higher the frequency the greater the energy dissipated. This dissipated energy results in heating of the dielectric. This effect can be a problem in some instances but also can be put to industrial use for the heating of materials. Domestic microwave ovens utilise this effect and operate at a frequency of 2.4 GHz. At such frequencies, water molecules have higher energy losses and thus food containing water is heated up uniformly and rapidly. Ceramics, paper and glass are not affected at such frequencies and so do not heat up.

## 4.7 Magnetic properties of solids

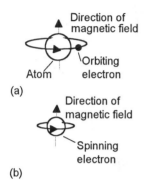

Figure 4.33 *(a) Orbiting electrons, (b) spinning electrons producing magnetic fields*

Figure 4.34 *Randomly orientated dipoles in a paramagnetic material*

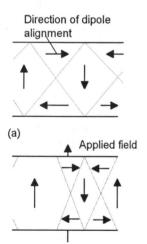

Figure 4.35 *Domains with (a) no applied field, (b) a field*

A current through a loop of wire, i.e. charges moving in a circular path, produces a magnetic field. We can consider atoms to have electrons in orbit around a nucleus, thus such orbiting electrons can be considered to generate magnetic fields (Figure 4.33(a)). Likewise an electron spinning on its axis can be considered to produce a magnetic field (Figure 4.33(b). Thus the electrons in atoms can be considered, by virtue of their movement, to make the atom behave like a tiny magnet. Such a tiny magnet is referred to as a *magnetic dipole*.

Materials can be grouped into three general categories:

1   *Diamagnetic materials*
    These have relative permeabilities slightly less than 1. Bismuth, copper, mercury and water are examples of such materials.

2   *Paramagnetic materials*
    These have relative permeabilities slight greater than 1. Aluminium and platinum are examples of such materials.

3   *Ferromagnetic and ferrimagnetic materials*
    These have relative permeabilities considerably greater than 1. Ferromagnetic materials are metals and ferrimagnetic materials ceramics.

We can explain diamagnetism as occurring with materials that have atoms for which the movements of the electrons in their orbits tend to produce magnetic effects which cancel each other out and there is thus no net magnetic field produced by an atom. Paramagnetic materials have atoms that behave as magnetic dipoles but, in the absence of a magnetic field, the dipoles are all randomly orientated (Figure 4.34) and thus the material shows no permanent magnetism.

Ferromagnetic materials have very high relative permeabilities. The atoms in such materials behave as permanent magnetic dipoles, with the way in which the atoms bond together in the solid resulting in neighbouring atomic dipoles aligning themselves all in the same direction. Their contributions thus add up. The dipoles in such a material are not generally all aligned in the same direction but in blocks within which they all are aligned in the same direction but neighbouring blocks may have different directions. Such regions in a material where magnetic dipoles are aligned in this way are called *domains*. Figure 4.35(a) shows the type of domain structure a material might have in the absence of an external magnetic field. The directions of the magnetic dipoles varies from domain to domain and may be completely random with the result that the material shows no permanent magnetism. When a magnetic field is applied, those domains with magnetic dipoles most nearly in the direction of the field grow in size at the expense of neighbouring domains (Figure 4.35(b)). Increasing the applied field increases their growth until eventually all the domains have dipoles in the direction of the applied field. The material is then said to be *saturated* and showing its greatest amount of magnetism. When the applied field is removed, many of the domains may remain orientated in the same direction and thus the material retains some residual magnetism.

If the temperature of a ferromagnetic material is raised, the thermal motion of the atoms increases. This motion tends to disrupt the alignment of magnetic dipoles in the domains. At a certain temperature, called the *Curie temperature*, the thermal motion completely overcomes the magnetic alignment and the ferromagnetic material looses its ferromagnetic properties.

## Problems

*Questions 1 to 17 have four answer options: A, B, C and D. Choose the correct answer from the answer options.*

**1** The weakest of the bonding forces between particles in a solid is:

A  The metallic bond.
B  The ionic bond.
C  The covalent bond.
D  The van der Waals bond.

**2** Decide whether each of these statements is TRUE (T) or FALSE (F).

Copper is a good conductor of electricity because:
(i)  It has free electrons.
(ii)  It forms a crystalline structure.

A  (i) T  (ii) T
B  (i) T  (ii) F
C  (i) F  (ii) T
D  (i) F  (ii) F

**3** Decide whether each of these statements is TRUE (T) or FALSE (F).

Polymers have a low stiffness because:
(i)  The carbon–carbon bonds in the polymer chain are weak.
(ii)  The bonds between polymer chains are strong.

A  (i) T  (ii) T
B  (i) T  (ii) F
C  (i) F  (ii) T
D  (i) F  (ii) F

**4** Decide whether each of these statements is TRUE (T) or FALSE (F).

The van der Waals bond is due to:
(i)  Attraction between ions.
(ii)  The sharing of electrons between atoms.

A  (i) T  (ii) T
B  (i) T  (ii) F
C  (i) F  (ii) T
D  (i) F  (ii) F

5  Decide whether each of these statements is TRUE (T) or FALSE (F).

The ionic bond between two atoms occurs as a result:
(i)   Electrons being shared by the atoms.
(ii)  Electrons moving from one atom to the other.

A  (i) T  (ii) T
B  (i) T  (ii) F
C  (i) F  (ii) T
D  (i) F  (ii) F

6  Decide whether each of these statements is TRUE (T) or FALSE (F).

The metallic bond in solids arises from the:
(i)   Attraction between the core ions.
(ii)  Attraction between the ion cores and the electrons.

A  (i) T  (ii) T
B  (i) T  (ii) F
C  (i) F  (ii) T
D  (i) F  (ii) F

7  Decide whether each of these statements is TRUE (T) or FALSE (F).

In a crystal, a unit cell is:
(i)   A group of atoms that forms a cubic arrangement.
(ii)  The smallest group of atoms that is regularly repeated.

A  (i) T  (ii) T
B  (i) T  (ii) F
C  (i) F  (ii) T
D  (i) F  (ii) F

8  Decide whether each of these statements is TRUE (T) or FALSE (F).

The molecules in a polymer are held together by:
(i)   Van der Waals or hydrogen bonds.
(ii)  Covalent bonds.

A  (i) T  (ii) T
B  (i) T  (ii) F
C  (i) F  (ii) T
D  (i) F  (ii) F

9  Decide whether each of these statements is TRUE (T) or FALSE (F).

Thermosetting polymers have:
(i)   A network structure.
(ii)  Very few cross-links between molecular chains.

A  (i) T  (ii) T
B  (i) T  (ii) F
C  (i) F  (ii) T
D  (i) F  (ii) F

10 Molecular chains in polymeric materials can be linear chains, chains with side groups or cross-linked chains forming a network structure.

Decide whether each of these statements is TRUE (T) or FALSE (F).

The rigidity of a polymeric material is increased if:
(i) The molecular chains have side groups.
(ii) The molecular chains form a network structure.

A (i) T (ii) T
B (i) T (ii) F
C (i) F (ii) T
D (i) F (ii) F

11 Decide whether each of these statements is TRUE (T) or FALSE (F).

The degree of crystallinity of a polymer is decreased if:
(i) The polymer chains have side groups.
(ii) The polymer chains are linear.

A (i) T (ii) T
B (i) T (ii) F
C (i) F (ii) T
D (i) F (ii) F

12 Decide whether each of these statements is TRUE (T) or FALSE (F).

Metals are good conductors of electricity because:
(i) They have free electrons.
(ii) They have positive ions.

A (i) T (ii) T
B (i) T (ii) F
C (i) F (ii) T
D (i) F (ii) F

13 Decide whether each of these statements is TRUE (T) or FALSE (F).

The capacitance of a parallel plate capacitor with air between its plates is increased if:
(i) The potential difference between the plates is increased.
(ii) A dielectric is introduced between the plates.

A (i) T (ii) T
B (i) T (ii) F
C (i) F (ii) T
D (i) F (ii) F

14 Decide whether each of these statements is TRUE (T) or FALSE (F).

Diamagnetic materials have:
(i) Relative permeabilities slightly below 1.
(ii) Atoms that do not behave like permanent magnetic dipoles.

A  (i) T  (ii) T
B  (i) T  (ii) F
C  (i) F  (ii) T
D  (i) F  (ii) F

15 Decide whether each of these statements is TRUE (T) or FALSE (F).

In a ferromagnetic material below its Curie temperature which is showing no permanent magnetism:
(i)  The domains have magnetic directions that are randomly orientated.
(ii) The directions of the dipoles within each domain are randomly orientated so that each domain has no resultant magnetism.

A  (i) T  (ii) T
B  (i) T  (ii) F
C  (i) F  (ii) T
D  (i) F  (ii) F

16 Decide whether each of these statements is TRUE (T) or FALSE (F).

Figure 4.36 shows how the resultant force between two atoms in two solids X and Y vary with their separation.
(i)  Solid X has larger atoms than solid Y.
(ii) Solid X has a larger value of Young's modulus than solid Y.

A  (i) T  (ii) T
B  (i) T  (ii) F
C  (i) F  (ii) T
D  (i) F  (ii) F

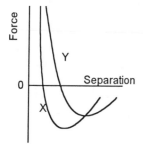

Figure 4.36 *Forces between particles*

17 When the atoms in a solid are separated by their equilibrium distance:

A  The force of attraction between the atoms is a maximum.
B  The force of repulsion between the atoms is zero.
C  The potential energy of the atoms is a minimum.
D  The potential energy of the atoms is zero.

18 Describe the basic structure of the following solids: (a) metals, (b) polymers, (c) ceramics.

19 Describe the basis of the following types of bonds: (a) metallic, (b) ionic, (c) covalent, (d) van der Waals.

20 Explain why the electrical conductivities of metals, insulators and semiconductors differ.

# 5 Structure and properties

## 5.1 Modifying materials

This chapter is about the relationship between the structure of materials and their properties and how the structure, and hence the properties, can be changed in order to produce materials which are more useful. The methods considered are:

1. Mixing elements or compounds to produce metallic alloys.
2. Subjecting metals to heat treatment in order to change the internal structure and hence the properties.
3. Changing the properties of polymers by altering the polymer chains.
4. Changing the properties of polymers by orientation of molecular chains.
5. Cross-linking polymer chains.
6. Adding materials to polymers to form plastics with specific properties.
7. Subjecting ceramics to heat treatment in order to change the internal structure and hence the properties.
8. Toughening glass by chemical treatment.
9. Combining materials to form composites.
10. Altering the electrical properties of semiconductors by adding donor atoms.

## 5.2 Alloys

An *alloy* is a metallic material made by mixing two or more elements or compounds to form a material with a specific composition. The everyday metallic objects around you will be made, almost invariably, from alloys rather then the pure metals themselves. Pure metals do not always have the appropriate combinations of properties needed; alloys can, however, be designed to have them. For example, the coins in your pocket are made of alloys. Coins need to be made of a relatively hard material that does not wear away rapidly, i.e. the coins have to have a 'life' of many years. Coins made of, say, pure copper would be very soft; not only would they suffer considerable wear but they would bend in your pocket. The copper-looking British coins are made of an alloy of copper with 2.5% by weight of zinc and 0.5% of tin, the term *coinage bronze* being used for the alloy. The silver-looking British coins are made of an alloy of copper with 25% by weight of nickel, the term cupro-nickel being used for the alloy.

Making alloys is rather like baking a cake. The basic ingredients of flour, sugar, fat, eggs and water are mixed together and then cooked. The result is a cake which has a texture and properties quite different from those of the individual ingredients. The type of cake produced depends on the relative amounts of the ingredients and the way it is cooked. In making alloys, the ingredients are mixed and heated and the resulting alloy can have properties quite different from those of the ingredients. The properties will depend on the relative amounts and nature of the ingredients as well as how they are 'baked'. An alloy is a particular mixture of components and so has a particular chemical composition, e.g. one carbon steel may be 99.0% iron combined with 1.0% carbon while another is 99.5% iron with 0.5% carbon.

Pure metals tend to be soft with high ductility, low tensile strength and low yield strength. Because of this they are rarely used in engineering. Alloying can produce harder materials with higher tensile strength, higher yield stress and a reduction in ductility. Such materials are generally more useful. There are, however, some circumstances in which the properties of pure metals are useful; for example, where high electrical conductivity is required (alloying reduces conductivity), where good corrosion resistance is required (alloying can result in less corrosion resistance), and where very high ductility is required.

We can think of the structure of alloys in terms of the constituents, say A and B, being mixed in the liquid state. Then when the mixture solidifies, there is the possibility that the solid alloy will have a crystal structure in which some of the atoms in the crystal structure of A have been replaced by atoms of B (Figure 5.1(a)). Alternatively, because there are spaces between the atoms of A in its crystal structure, some atoms of A, if small enough, might lodge in these spaces (Figure 5.1(b)). Another possibility is that elements A and B combines to form a chemical compound. With a compound, there will be a particular structure for that compound with atoms of A and B assuming specific positions, rather than just popping into any gap. Another possibility is that, when the liquid mixture cools, A and B separate out, with B forming its own crystal structure independent of A. The structure then becomes a mixture of two types of crystals.

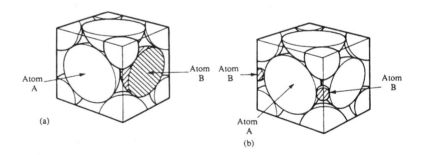

Figure 5.1 *Possible forms of alloys*

### 5.2.1 Ferrous alloys

Pure iron is a relatively soft material and is hardly ever used. Alloys of iron with carbon are, however, very widely used, the term *ferrous alloys* being used for such alloys. Pure iron at room temperature exists as a body-centred cubic structure, this being commonly referred to as *ferrite*. This form continues up to 912° C. At this temperature, the structure changes to face-centred cubic, known as *austenite*. But what happens to the structures when we have carbon present?

Iron atoms have a diameter of 0.256 nm (1 nanometre = $10^{-9}$ m), carbon atoms are much smaller with a diameter of 0.154 nm. A face-centred cubic structure is a more open structure than the body-centred structure. Thus, the face-centred structure of austenite has voids that can accommodate spheres up to 0.104 nm in diameter while the body-centred cubic structure has voids between the atoms which can accomodate spheres up to 0.070 nm in diameter. Thus, carbon atoms can be more easily accommodated within austenite, without severe distortion of the lattice, than in ferrite. Austenite can take up to 2.0% of carbon while ferrite can only take 0.2%. Thus when iron containing carbon is cooled from the austenite state to the ferrite state, there is a reduction in the amount of carbon that can be accommodated within the iron and so some of the carbon atoms come out of the crystals and form a compound, another crystal structure, between iron and carbon called *cementite*. Cementite is hard and brittle. The result can be a structure consisting of purely ferrite grains mixed with grains that have a laminated structure of  ferrite and cementite. Such a laminated structure is termed *pearlite*. Pure cementite is harder than pearlite, which in turn is harder than pure ferrite.

The percentage of carbon alloyed with iron has a profound effect on the properties of the alloy. The terms used for the alloys produced with different percentages of carbon are:

Wrought iron  0 to 0.05% carbon
Steel         0.05 to 2.0% carbon
Cast iron     2.0 to 4.5% carbon

The term *carbon steel* is used for those steels in which essentially just iron and carbon are present. The term *alloy steel* is used where other elements are included.

### 5.2.2 Carbon steels

Carbon steels are grouped according to their carbon content with the designations being roughly:

Mild steel            0.10 to 0.25% carbon
Medium-carbon steel  0.20% to 0.50% carbon
High-carbon steel     More than 0.50% carbon

Mild steel has a structure consisting predominantly of ferrite, medium-carbon steels tend to have about equal amounts of ferrite and pearlite, while high-carbon steels have predominantly pearlite with some free cementite occurring at high-carbon contents.

Figure 5.2 shows how the mechanical properties of carbon steels depend on the percentage of carbon. Increasing the percentage of carbon, within the range considered, increases the amount of pearlite at the expense of the softer ferrite, and hence:

1   Increases the tensile strength.
2   Increases the hardness.
3   Decreases the percentage elongation.
4   Decreases the impact strength.

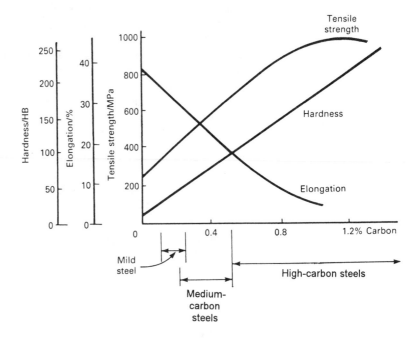

Figure 5.2  *Properties of carbon steels*

Mild steel is a general purpose steel and is used where hardness and tensile strength are not the most important requirements but ductility is often required. Typical applications are sections for use as joists in buildings, bodywork for cars and ships, screws, nails, wire. Medium-carbon steel is used for agricultural tools, fasteners, dynamo and motor shafts, crankshafts, connecting rods, gears. With such steels, the lower ductility puts a limit on the types of processes that can be used. Medium-carbon steels are capable of being quenched and tempered to develop reasonable toughness with strength (see Section 5.3). High-carbon steel is used for withstanding wear, where hardness is a more necessary require-ment than ductility. It is used for machine tools, saws, hammers, cold chisels, punches, axes, dies, taps, drills, razors. The main use of high-carbon steel is as a tool steel. High-carbon steels are usually quenched and tempered at about 250°C to develop their high strength with some slight ductility (see Section 5.3).

**Example**

A pickaxe head may be made of a high-carbon steel. Why high-carbon rather than mild steel?

High-carbon steel is a harder, stronger material than mild steel. The higher ductility of mild steel is not required in this situation.

**Example**

A carbon steel is said to have a predominantly ferrite structure. What properties could be expected?

A ferrite structure gives a soft, ductile, low strength structure. See Figure 5.2.

### 5.2.3 Non-ferrous alloys

The term *non-ferrous alloy* is used for all alloys where iron is not the main constituent. Non-ferrous alloys include alloys of aluminium, copper, magnesium, etc. The following are some of the general properties and uses of non-ferrous alloys in common use in engineering:

| | |
|---|---|
| Aluminium alloys | Low density, good electrical and thermal conductivity, high corrosion resistance. Tensile strengths of the order of 150 to 400 MPa, tensile modulus about 70 GPa. Used for metal boxes, cooking utensils, aircraft bodywork and parts. |
| Copper alloys | Good electrical and thermal conductivity, high corrosion resistance, tensile strengths about 180 to 300 MPa, tensile modulus about 20 to 28 GPa. Used for pump and valve parts, coins, instrument parts, springs, screws. |
| Magnesium alloys | Low density, good electrical and thermal conductivity. Tensile strengths of the order of 250 MPa and tensile modulus about 40 GPa. Used as castings and forgings in the aircraft industry where weight is an important consideration. |
| Nickel alloys | Good electrical and thermal conductivity, high corrosion resistance, can be used at high temperatures. Tensile strengths between about 350 and 1400 MPa, tensile modulus about 220 GPa. Used for pipes and containers in the chemical industry where high resistance to corrosive atmospheres is required, food-processing equipment, gas turbine parts. |

Titanium alloys     Low density, high strength, high corrosion resistance, can be used at high temperatures. Tensile strengths of the order of 1000 MPa, tensile modulus about 110 GPa. Used in aircraft for compressor discs, blades and casings, in chemical plant where high resistance to corrosive atmospheres is required.

Zinc alloys         Low melting points, good electrical and thermal conductivities, high corrosion resistance, tensile strength about 300 MPa, tensile modulus about 100 GPa. Used for car door handles, toys, car carburettor bodies – components that in general are produced by pouring the liquid metal into dies.

As an example of a non-ferrous alloy, consider copper alloys. Pure copper is a soft material with low tensile strength. For many engineering purposes it is alloyed with other metals. The exception is where high electrical conductivity is required. Pure copper has a better conductivity than the alloys. The following indicates the names given to the various types of copper alloys:

| | |
|---|---|
| Copper with zinc | Brasses |
| Copper with tin | Bronzes |
| Copper with tin and phosphorus | Phosphor bronzes |
| Copper with tin and zinc | Gunmetals |
| Copper with aluminium | Aluminium bronzes |
| Copper with nickel | Cupro-nickels |
| Copper with zinc and nickel | Nickel silvers |
| Copper and silicon | Silicon bronze |
| Copper and beryllium | Beryllium bronze |

Figure 5.3 shows how the percentage of zinc included with copper to form brasses affects the mechanical properties. Brasses with between 5 and 20% zinc are called gilding metals and, as the name implies, are used for architectural and decorative items to give a 'gilded' or golden colour. It is used for coins, medals and decorative architectural items. Cartridge brass is copper with 30% zinc. One of its main uses is for cartridge cases, items which require high ductility for the deep drawing process used to make them. The term *basis brass* is used for copper with 37% zinc. This is a good alloy for general used with cold working processes and is used for fasteners and electrical connectors. It does not have the high ductility of those brasses with less zinc. Copper with 40% zinc is called Muntz metal. This metal, often has some lead added to improve its machining properties, and is used for decorative brassware, heat condenser plates and a range of items produced by forging and pressing

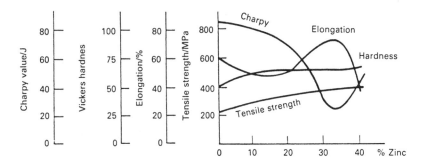

Figure 5.3 *Properties of brasses*

The changes in the properties of brasses when the amount of zinc is changed arises from changes in the structure. Brasses with between 0 and 35% zinc form one type of structure, termed alpha, and between 5% and 45% there is a mixture of this alpha structure and another structure termed beta. It is this change in structure, i.e. the way the atoms of copper and zinc are packed together, that is responsible for the abrupt changes in properties of brass at 35% zinc.

**Example**

What alloys might be considered if a metallic material is required with high corrosion resistance and high tensile strength?

Titanium alloys have, in general, good corrosion resistance and high tensile strengths.

## 5.3 Models for metal structure

A simple way we can think of the atoms in a metal is as though they were an orderly array of spheres tethered to each other by springs (see Figure 4.1). When forces are applied to stretch the material, then the springs are stretched and exert an attractive force pulling the atoms back to their original positions. When forces are applied to compress the material, then the springs are compressed and exert repulsive forces which push the atoms back to their original positions. See Section 4.2 for a discussion of these forces. This then represents a simple model for a metal.

A simple theory to explain the elastic and plastic behaviour of metals when stretched is the *block slip theory*. Consider a block of atoms in the form suggested by the above model. In the absence of any externally applied forces the atoms are all in their equilibrium positions (Figure 5.4(a)). When stress is applied to a metal, then we can consider that the block of atoms is at such an angle to the forces that the situation becomes as shown in Figure 5.4(b). Elastic strain occurs when atoms become displaced from their equilibrium positions by an applied stress but are able to spring back to them when the stress is removed. However, if the stress is high enough then yielding occurs and blocks of atoms slip (Figure 5.4(c)).

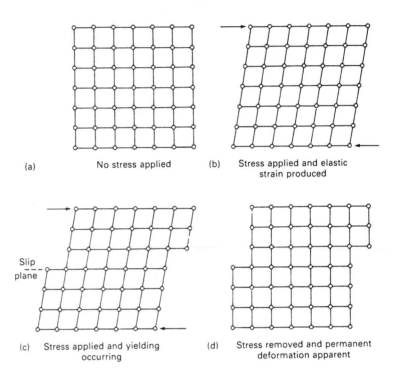

(a)   No stress applied

(b)   Stress applied and elastic strain produced

Slip plane

(c)   Stress applied and yielding occurring

(d)   Stress removed and permanent deformation apparent

Figure 5.4 *Block slip model*

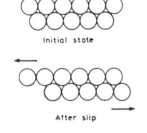

Initial state

After slip

Figure 5.5 *Slip of atomic planes*

When the stress is removed the atoms spring back to equilibrium positions but for some atoms these are new positions and permanent deformation has been produced (Figure 5.4(d)). The plane along which atoms slip is called the *slip plane*. In terms of our model of a crystal as a pile of stacked spheres, we can consider that slip is when an entire row of spheres is pushed sufficiently to all move each atom along one position, as illustrated in Figure 5.5.

On this model of slip, we can only have slip within an orderly arrangement of spheres, i.e. within a grain. Slip planes cannot cross over from one grain to another, the disorderly arrangement at the grain boundary does not allow it. Slip will only occur in those grains that have atomic planes at suitable angles to the applied forces. Thus a metal having big grains can have more slippage than one having a larger number of small grains. A simple model we might use is to consider soldiers on parade in orderly ranks, i.e. all the soldiers in one large 'grain'. For the movement of one soldier in the back rank to step forward, then all the soldiers in that line step forward and there is 'slip' and a large amount of movement. The analogy with the small grain metal is of a crowd of people in which locally there might be some small pockets of orderly arrangement but certainly not order over a large segment of the crowd. When one person moves, then there might be some local 'slip' as people move but there is no overall movement of the crowd. Thus, with metals, the bigger the grains, the greater the amount of plastic deformation that might be expected. A fine grain structure will have less slippage and so show less plastic deformation, i.e. be less ductile. A brittle material is thus one in which each slip process is confined to a short run in the metal and not allowed to spread. A ductile material is one in which the slip process is not confined to a short run in the metal and does spread over a large part of the metal.

While there can be considered to be many planes of atoms in a crystal, slip is found to only occur between the planes with the closest packing of atoms. This is because the atoms are close enough to permit changes of positions more easily than when further apart. Figure 5.6 illustrates this concept of close-packed planes, the lines indicating the planes with the highest density of atoms per unit length. If you draw other lines through atoms, you will find less atoms per unit length. The number of such high density planes along which slip can occur depends on the form of structure of the crystal. The body-centred cubic structure has many such slip planes, the face-centred cubic less and the hexagonal close-packed structure even less. Thus metals which have a hexagonal close-packed structure tend to be harder and more brittle than metals with the face-centred cubic structure, while the body-centred cubic structure is likely to be the least hard and most ductile metal.

The above is just a simple model of what happens when metals are stretched. The model needs modification to do more than give a simple idea of what happens. It has assumed that the arrangement of atoms within a grain is perfectly orderly. In reality this is not the case and there are some atoms in the wrong places, the term *dislocations* being used. Thus, within a grain, we might have the situation shown in Figure 5.7(a). When stress is applied, the dislocation moves though the array of atoms, as illustrated in Figure 5.7(b)(c)(d) so the slip takes place atom-by-atom rather than the wholesale movement of one plane of atoms past another.

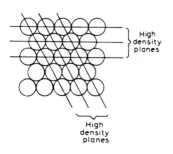

Figure 5.6 *High density planes*

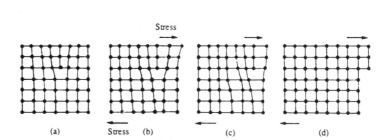

Figure 5.7 *Movement of a dislocation under the action of stress*

When the movement of a dislocation through a metal brings it up to another dislocation, then they can either cancel each other out or hinder further movement of dislocations. Figure 5.8 shows what can happen when two dislocations come close to each other. Each dislocation has the atoms on one side of the slip plane in compression and on the other side in tension. When two compression regions come close together, the forces between the atoms result in the dislocations repelling each other. In general, the more dislocations a metal has, the more the dislocations get in the way of each other and so the more difficult it is for the dislocations to move through the metal and hence slip to occur. Thus the greater the number of dislocations, the greater the stress needed to produce yielding.

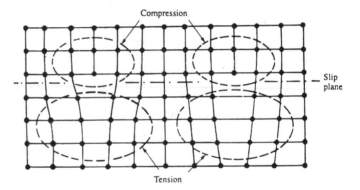

Figure 5.8 *Dislocations repelling each other*

Dislocations are produced as a result of missing atoms, atoms being displaced from their correct positions, and foreign atoms being present and distorting the orderly packing of atoms. Cold working (see Section 5.4), which distorts grains, results in an increase in dislocations because it displaces atoms from their correct positions. Foreign atoms may be present as a result of a deliberate alloying process. Thus alloying, in increasing the number of dislocations and making it more difficult for dislocations to move through a material, increases the yield stress. Another method of increasing the yield stress is to cause small particles to precipitate out from an alloy. Such particles may cause more dislocations or block the movement of dislocations through the material. Such a process is called *precipitation hardening*.

The above is just an indication of how the properties of metals can be explained in terms of their structure. You can carry out simple experiments to gain an understanding of how dislocations behave by the use of bubble rafts (see Section 4.3.1). Make a bubble raft containing dislocations between pieces of wire and study what happens when the wires are pushed towards each other and squeeze the bubble raft. The dislocations will be seen running along rows of bubbles. If the number of dislocations are increased, by perhaps bursting some of the bubbles with a hot wire, then the ease with which the bubble raft can be compressed can be examined. Increasing the number of dislocations makes it harder for slip to transmit through the raft. You might also try introducing a few larger bubbles into the raft and see their effect on the movement of dislocations.

## 5.4 Working and heat-treating metals

Suppose you were to take a carbon steel test piece and perform a tensile test on it. You may, for instance, find that the material showed a yield stress of 430 MPa. If the stress is continued beyond this point but the stress released before the material breaks, then a permanent deformation will be found to have occurred. Figure 5.9(a) illustrates this sequence of events and indicates a permanent deformation of a strain of 0.015. Now suppose we start the tensile test all over again and stretch the material. This time the yield stress is not 430 MPa but 550 MPa. The material has a much higher yield stress (Figure 5.9(b)). It is not only the yield stress that has increased, the tensile strength has increased, the percentage elongation has decreased, and the hardness has increased. After plastic deformation, the mechanical properties of the steel have changed.

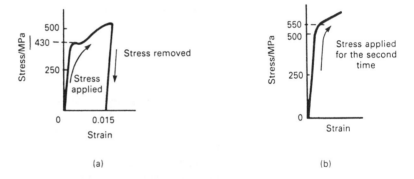

Figure 5.9 *Stretching a carbon steel*

The material is said to have been subject to *cold working* and the above are typical of the changes that occur. The term cold working is used when plastic deformation is produced at a temperature that is not high enough itself to produce changes. The term *work hardening* is often used since the cold working has resulted in the material becoming harder. Such work hardening occurs when, for example, a metal sheet is being rolled to reduce its thickness. The rolling makes the material harder and more brittle, and thus there is a limit to the amount of reduction in thickness that is possible without some form of heat treatment to make the metal soft and ductile again.

We can offer an explanation for these effects of cold working on the properties of metals by considering that the plastic deformation results in slip and distortion, with the result that the number of dislocations of atoms within grains increases. It is then more difficult to produce further slip, hence the increase in brittleness.

### 5.4.1 Heat-treating cold-worked metals

Cold working of metals results in changes in mechanical properties with yield stress, tensile strength and hardness increasing, and percentage elongation decreasing. These changes can be reversed by suitable heat treatment. Cold-worked metals generally have deformed grains, with a high density of dislocations within the grains. Such dislocations lead to internal stresses within grains. When such a metal is heated, then there is some slight rearrangement of atoms within the grains and a consequent reduction in internal stresses. This process is known as *recovery*.

When a cold-worked metal is heated above about $0.3T_m$, where $T_m$ is the melting point of the material on the kelvin scale of temperature, then there is a marked reduction in tensile strength, hardness and an increase in percentage elongation. What is happening is that the material is beginning to *recrystallise*. New grains start to grow. The temperature at which recrystallisation starts is called the *recrystallisation temperature*. For pure metals, it tends to be about 0.3 to $0.5T_m$. Thus, aluminium, which has a melting point of 933 K, has a recrystallisation temperature of 423 K, about $0.45T_m$. Iron with a melting point of 1356 K has a recrystallisation temperature of 473 K, about $0.35T_m$. As the temperature is further increased, so the crystals start to grow until they have completely replaced the original distorted cold worked structure. Figure 5.10 illustrates the sequence of events.

The term *annealing* is used for the heat treatment process that involves heating the material to above the recrystallisation temperature and cooling it slow enough for the large new grains to persist and so give a material with more ductile properties. However, if iron is not cooled slowly but *quenched* by dropping it in cold water, the resulting material is much harder and more brittle. This is because, slow cooling allows time for the excess carbon in the austenite structure to escape, fast cooling does not. The result of fast cooling is distortion of the crystal structure in order to accommodate the excess carbon and a new crystal structure, called *martensite*. Martensite is very hard and brittle. Ductility can be restored by heating the alloy to a temperature at which carbon atoms can diffuse out of the martensite structure. This process is called *tempering*. The temperature and time for which the alloy is heated will determine the amount of carbon that diffuses out of the martensite structure and hence the hardness and ductility of the resulting material.

*Heat treatment* can be defined as the controlled heating and cooling of metals in the solid state for the purpose of altering their properties. A heat treatment cycle consists normally of three parts:

1   Heating the metal to the required temperature for the changes in structure within the material to occur.

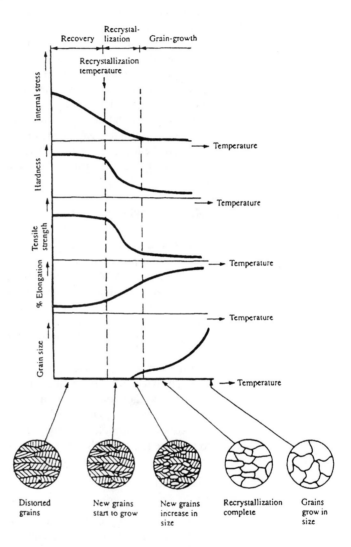

Figure 5.10 *The effect of heat treatment on cold-worked metals*

2 Holding the material at that temperature for long enough for the entire material to reach the required temperature and the structural changes to occur through the entire material.
3 Cooling, with the rate of cooling being controlled since it affects the resulting structure and hence the properties of the material.

Annealing is the heat treatment used to make a metal softer and more ductile. Quenching is the heat treatment used to make a material hard and brittle. This is generally linked with tempering, the process used to restore a quenched material to an acceptable level of hardness and ductility.

You can see the effects of heat treatment with a simple experiment involving steel sewing needles. A note of caution: eye protection needs to be worn during this experiment. Bend a steel sewing needle. You should find that it will show significant bending before it breaks; a property required of sewing needles if they are to be used in sewing without breaking. Now heat a sewing needle in a Bunsen flame and then allow it to cool slowly, i.e. an annealing treatment. The needle now bends very easily, being very soft and ductile. Now heat a sewing needle in the Bunsen flame and the drop it into cold water, i.e. a quenching treatment. The needle will not now bend easily and is hard and brittle. You can temper such a quenched needle by warming it for a little while.

### 5.4.2 Hot working

Cold working involves plastically deforming materials at temperatures that are below the recrystallisation temperature. The result is a harder, less ductile material with deformed grains. *Hot working* involves deforming a material at a temperature greater than the recrystallisation temperature. Then, as soon as a grain becomes deformed, it recrystallises. No hardening thus occurs and the working can be continued without any difficulty.

A disadvantage of hot working is that oxidisation of the metal surfaces occurs. Cold working does not have this problem. Another disadvantage is that the material will have comparatively low values of hardness and tensile strength, with high percentage elongation. A combination of hot and cold working is thus often used in a particular shaping process. The first operation involving large amounts of plastic deformation is carried out by hot working. After cleaning the surfaces of the metal, it is then cold worked to increase the strength and hardness and give a good surface finish.

### Example

Lead has a melting point of 327°C. Will the product be work hardened if it is made by extruding at room temperature? Extrusion is a process rather similar to the squeezing of toothpaste out of a tube, the metal being squeezed out through a nozzle and taking the shape dictated by that of the nozzle.

The melting point of lead is about 600 K. This would mean the extrusion at about 300 K is at about $0.5T_m$. In other words, the extrusion is taking place at about the recrystallisation temperature. The process is likely to just about be a hot-working process and so there would be no work hardening.

**5.5 Polymers**     Thermoplastics consist of polymers with long chain molecules that are either linear chains or long chains with small branches. Linear chains have no side branches or cross-links with other chains. Because of this they can easily move past each other. If, however, the chain has branches, then there is a reduction in the ease with which chains can be made to move past each other. This shows itself in the material being more stiff, i.e. less strain produced for a given stress. Making cross-links between chains makes it even more difficult to stretch the material. Thermosets are polymers with considerable cross-links and consist of a networked of linked atoms. The following is a brief discussion indicating how changes to a basic polymer chain can produce changes in properties.

*Polyethylene* can be processed to have linear chains (see Figure 4.21(a)). The chains have a core of carbon atoms with hydrogen atoms attached, essentially an almost endless repetition of $-CH_2-$ units. The hydrogen atoms are small and bed into the carbon chain to give a very smooth, linear, chain. There is freedom for the chain to twist about any C–C bond and so the chain is flexible. Scaled up, the polyethylene chain is rather like a piece of string about 2 m long. The forces between the chains are due to the weak van der Waals bonding. The result of having such smooth, linear chains is that a high degree of crystallinity is possible. However, we can add knobs and side branches to such a basic chain and so considerably alter the properties of the solid polymer.

*Polypropylene* has some of the hydrogen atoms replaced by $CH_3$ groups regularly arranged along the chain (Figure 5.11). The result is knobbly chains and a material that is more rigid and stronger than polyethylene in its linear form. The crystallinity is about 60%; lower than that of linear polyethylene. Polypropylene is used for crates, containers, fans, car fascia panels, tops of washing machines, cabinets for radios and televisions, toys chair shells.

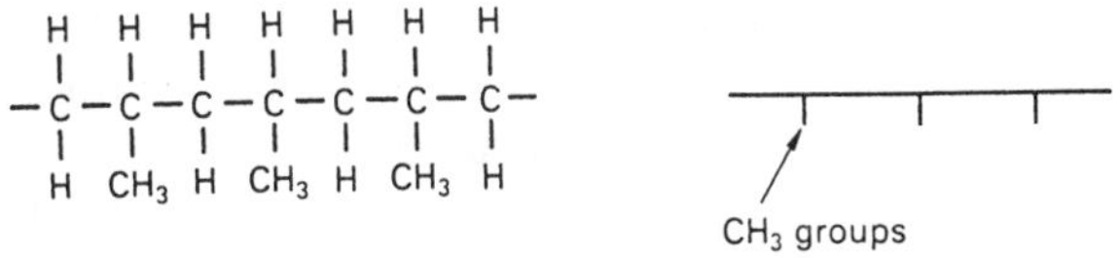

Figure 5.11 *Polypropylene*

*Polyvinyl chloride* (PVC) has a linear chain but has 'bulky' atoms, chlorine atoms, replacing some hydrogen atoms on the chain (Figure 5.12). Because of this the chain very knobbly and it is difficult to pack such chains in an orderly manner. The result is a mainly amorphous material. When used without a plasticiser (see Section 5.5.2) it is a rigid and relatively hard material. It is widely used for piping, but not for hot water as it has a maximum service temperature of 70°C. Above that temperature it softens too much. Most PVC products are, however, made with a plasticiser incorporated with the polymer. The amount of plasticiser is likely to be between about 5 to 50% of the plastic, the more plasticiser added, the greater the degree of flexibility. Plasticised PVC is used for the fabric of plastic raincoats, bottles, shoe soles, garden hose piping, gaskets and inflatable toys. All forms of PVC have good chemical resistance, though not as good as polyethylene.

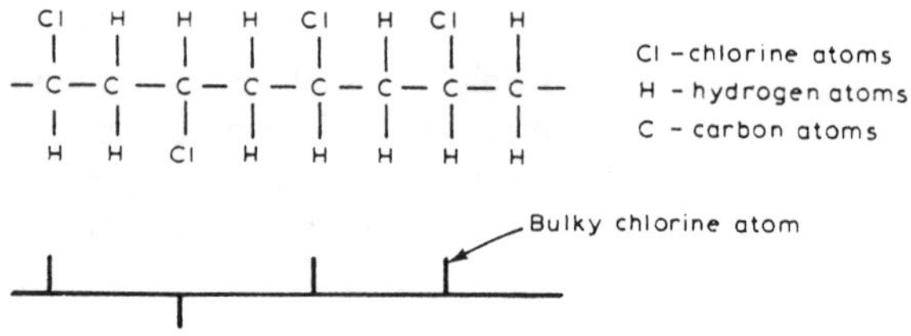

Figure 5.12　*PVC*

An alternative to putting knobs or branches on to a $-CH_2-$ chain is to make the chain stiffer by incorporating blocks in the backbone of the chain. An example of such a polymer is *polyethylene terephthalate (PET)*. This incorporates a six-carbon (benzene) ring in the backbone. This ring structure will not twist like the C–C bond and so the chain is stiffer. The polymer is widely used for the plastic bottles used for Coca-Cola and other drinks. It is also used as fibres for clothing, being known then as Terylene or Dacron.

The polymers discussed above are essentially linear polymer chains with the weak van der Waals bonding providing the forces between chains. In such materials, the individual chain molecules are distinct and capable of being separated. Such materials are termed *thermoplastics*. Another possibility with polymers is to have structures consisting of networks of chemical bonds. Such materials are generally *thermosets* or *elastomers*.

Many elastomers consist of linear chains linked by small molecules. *Natural rubber* is an example of such a material. When rubber is *vulcanised*, molecules of sulphur form cross-links between chains. The amount of sulphur added determines the amount of cross-links and hence the properties of the rubber. The greater the number of cross-links, the harder it is to stretch the rubber. The rubber of a rubber band has typically about one sulphur cross-link every few hundred carbon atoms.

Many thermosets consist of small molecules which are linked together to form a network. This gives a very stiff material. *Phenolics* give highly cross-linked polymers, thermosets. *Phenol formaldehyde* was the first synthetic plastic and is known as *Bakelite*. The polymer is opaque and initially light in colour. It does, however, darken with time and so is always mixed with dark pigments to give a dark coloured material. It is supplied in the form of a moulding powder that includes the polymer, fillers and other additives such as pigments. When this moulding powder is heated in a mould the cross-linked polymer chain is produced. Phenol formaldehyde mouldings are used for electrical plugs and sockets, switches, door knobs and handles, camera bodies and ash-trays. Composite materials involving the polymer being used with paper or an open-weave fabric, e.g. a glass fibre fabric, are used for gears, bearings and electrical insulation parts (see Section 5.8).

### Example

What structural changes could be used to make a polymer material consisting of linear chains of $-CH_2-$ groups a stiffer material?

The stiffness can be increased by: replacing some of the hydrogen atoms by bulky atoms or groups of atoms, introducing side branches to the chain, replacing some of the carbon atoms by groups of atoms, or introducing cross-links between polymer chains.

### 5.5.1 Stretching polymers

Consider what happens with a crystalline thermoplastic when it is stretched. Figure 5.13 shows the typical form of stress–strain graph.

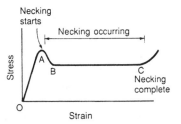

Figure 5.13 *Stress–strain graph for a crystalline polymer*

When stress is applied, the first thing that begins to happen is that there is some movement of folded chains past each other. However, when point A is reached the polymer chains start to unfold to give a material with the chains all lined up along the direction of the forces stretching the material. The material shows this by starting to exhibit *necking* (Figure 5.14), i.e. a section of the material suddenly shows a marked contraction in its cross-section. As the stress is further increased, the necking spreads along the material with more and more chains unfolding. Eventually, when the entire material is at the necked stage, all the chains have lined up. The material in this state behaves differently to earlier in the stress–strain graph, the material being said to be *cold drawn*. Such a material has, as a result of the orientation of the molecular chains, different properties to the undrawn material. The material is stiffer, i.e. tensile modulus is higher, and stronger. Typically, with polyethylene, the tensile modulus increases from about 1 GPa to 10 GPa and the tensile strength from about 30 MPa to perhaps 200 MPa. However, the percentage elongation is reduced, typically from about a few hundred per cent to less than 10%.

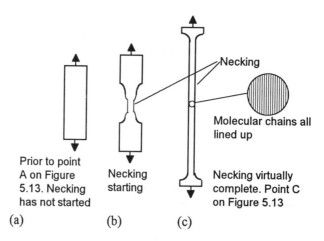

Figure 5.14  *Necking in a polymer*

The above sequence of events only tends to occur if the material is stretched slowly and sufficient time elapses for the molecular chains to unfold. If a high strain rate is used, the material is likely to break without the chains all becoming lined up. The plastic used for making polythene bags is a crystalline polymer. Try cutting a strip of polythene from such a bag and pulling it between your hands and see the necking develop with low rates of strain. Try quickly breaking a strip of polyethylene before orientating the molecules and then another strip after it has been stretched and the molecules orientated, the difference in tensile strength should be apparent.

The above is a consideration of what happens when crystalline polymers are stretched. Now consider what happens with amorphous polymers. Below the glass transition temperature, the polymer is glass-like and rather stiff and brittle. This is because, when so cold, no chains or parts of chains can move. If the temperature is increased to above the glass transition temperature, the material behaves in a rubbery fashion. This polymer is then very flexible, i.e. a much lower value for Young's modulus, and is able to withstand large and recoverable strains, i.e. just like a rubber band. This is because there is now sufficient thermal energy being supplied for not only side groups on chains to be able to rotate but also entire segments of the chain also rotate and move. Figure 5.15 shows how Young's modulus varies for such a material. Note that the modulus is plotted on a scale where each scale marking sees the modulus value increase by a power of ten.

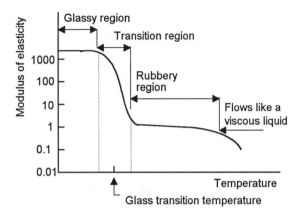

Figure 5.15 *Effect of temperature on the modulus of elasticity of an amorphous polymer*

Elastomers are polymers that at room temperature are above their glass transition temperatures and so exhibit rubbery behaviour. However, if you cool a rubber sufficiently it becomes brittle and shows glassy behaviour. Ordinary rubber tubing if cooled in liquid nitrogen shows such behaviour. Most polymers become rubbery at some temperature, the exception being heavily cross-linked thermosets, which decompose before they reach their glass transition temperatures.

### 5.5.2 Additives

The term *plastic* is commonly used to describe materials based on polymers. Such materials, however, invariably contain other substances that are added to the polymers to give the required properties. Since some polymers are damaged by ultraviolet radiation, protracted exposure to the sun can lead to a deterioration of mechanical properties. An ultraviolet absorber is thus often added to the polymer, such an additive being called a *stabiliser*. Carbon black is often used for this purpose. *Plasticisers* are added to the polymers to make it more flexible. In one form this may be a liquid which is dispersed throughout the solid, filling the space between the polymer chains and acting like a lubricant and permitting the chains to more easily slide past each other. *Flame retardants* may be added to improve fire-resistant properties, pigments and dyes to give colour. The properties and cost of a plastic can be markedly affected by the addition of substances termed *fillers*. Since fillers are generally cheaper than the polymer, the overall cost of the plastic is reduced. Up to 80% of a plastic may be filler. Examples of fillers are glass fibres to increase the tensile strength and impact strength, mica to improve electrical resistance, graphite to reduce friction, wood flour to increase tensile strength. One form of additive used is a gas to give foamed or expanded plastics. Expanded polystyrene is used as a lightweight packaging material, foamed polyurethane as a filling for upholstery.

## 5.6 Ceramics and glasses

Ceramics, such as diamond, corundum ($Al_2O_3$) and silicon carbide (SiC), are the hardest of solids, much harder than any metal. In a hardness test an indenter is pressed into the surface of the material, the size of the resulting deformation being a measure of the hardness. Thus a hard material is one that requires a higher stress to give yielding and so deformation. Ceramics have high yield stresses. Ceramics are, however, brittle and not tough. Such properties can be explained in terms of their structure.

Crystalline ceramics can be considered in terms of an orderly packing of spheres; the model we can use for crystalline ceramics is the dislocation model, discussed earlier in this chapter for metals. Unlike metals where the bonds are metallic bonds, ceramics are networks of ionic or covalent bonds. When a dislocation moves through a ceramic material, it has to break and reform these bonds as it moves. Because the bonds are stronger, this is more difficult than with a metal. In the case of the ionic bonds we also have movement only feasible along planes that do not bring ions of the same sign into juxtaposition (Figure 5.16). There are far fewer such slip planes than with a metal, thus it is much more difficult for dislocations to move through such a material. Thus ceramic materials are hard and brittle, because dislocations cannot readily move through them. This hardness is exploited in the use of diamond for cutting hard materials and in grinding wheels where ceramic particles are bonded to the wheel.

Figure 5.16 *Slip with an ionic crystal*

We can explain the lack of toughness of ceramics by using the same model. Consider what happens when we have a crack in a material. When subject to stress, the stress at the tip of a crack is intensified. With a ductile material, the stress at the tip of the crack will result in the material at the point yielding and plastic deformation occurring. The effect of this is to blunt the tip of the crack and so reduce the stress concentration. With a brittle material, no significant plastic deformation occurs and so the crack is not blunted and the high stress continues, with the result that the material fails. Ceramics have low toughness because they are brittle materials. In addition, because of the methods used for the production of ceramic products, they generally contain small cracks and flaws. The consequence of having these small cracks and the low toughness is that ceramic materials readily fail in tension.

Non-crystalline ceramics and glasses are also hard, brittle and with low toughness. With such materials there is no orderly array as with crystalline ceramics. We do, however, have a network of atoms or ions tightly bound together by ionic or covalent bonds. For yielding to occur we must have movement of atoms or molecules and, since they cannot easily move any distance, the material does not easily yield. Thus the material is hard and brittle. The low toughness arises, as described above for crystalline ceramics, because of this brittleness.

Glass when freshly produced has a much higher tensile strength than after some exposure to the environment. This is because initially there are few surface cracks or flaws to act as centres for crack propagation. Exposure to the environment rapidly results in such cracks appearing and so a large reduction in strength. Freshly produced glass fibres when used in composite materials and protected from the environment can have very high strengths and are a useful form of reinforcement.

### 5.7.1 Toughening ceramics

Ceramics are rather weak materials in tension, cracks readily propagating through them. The following discussion illustrates how two particular types of ceramic can be toughened, namely glass and concrete.

With glass, it is generally surface flaws that act as centres for crack propagation under tensile stress. Tensile stress acts in such a way as to pull open cracks, compressive stress would close them. Thus compressive stress does not cause cracks to propagate and so to prevent glass fracturing we have to keep the surfaces in compression. We can do this by using a treatment that builds compressive stresses in to the material surface (these are generally referred to as *residual stresses*), so giving a stress situation across the thickness of the material of the form shown in Figure 5.17. When a tensile stress is applied to the material, it has to exceed the surface internal compressive stresses before the surface is in tension. As a result, a surface with such surface compressive stresses can withstand the application of a larger tensile stress without fracturing than one with no internal compressive stresses. Glass with such surface internal compressive stresses is called *toughened glass* and is used for car windscreens.

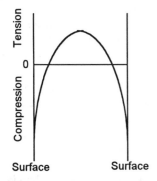

Figure 5.17 *Internal stresses with toughened glass*

There are two types of method used for producing surface internal compressive stresses. *Thermal toughening* involves heating a sheet of glass to a uniform temperature close to its softening point. The surfaces are then cooled uniformly and rapidly by cold air jets. The thermal contraction of the surfaces occurs freely because the interior of the glass is still hot and soft. The result is cooler, hardened, surface layers with a hotter, softer, core. When the core begins to cool, it can only contract by pulling against the already hardened surface layers. This results in the surface layers being put in compression and the core in tension. *Chemical toughening* involves some of the ions in the surface layers of the glass being replaced by larger dia-meter ions, e.g. sodium ions with a diameter of 0.196 nm might be replaced by potassium ions with a diameter of 0.266 nm. The larger ions would cause the surface layers to expand, but this expansion is resisted by the untreated core. The result is a surface in compression and a core in tension.

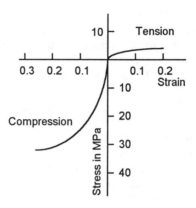

Figure 5.18  *Stress–strain graph for concrete*

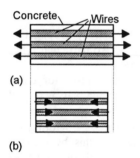

Figure 5.19  *Prestressed concrete*

Concrete as a structural material has the disadvantage of low strength in tension. However, it has high strength in compression. Figure 5.18 shows a typical stress–strain graph. Concrete is about eight times stronger in compression than tension. *Prestressed concrete* uses this to give a much stronger concrete product. Concrete in a beam is allowed to solidify round steel wires (Figure 5.19(a)) which are held in a state of tension. When the concrete has set, the tensile stress on the wires is removed and the wires contract, forcing the concrete into compression. The concrete has thus internal compressive stresses. When it is subject to tensile stresses, these have to exceed the internal compressive stresses before the concrete goes into tension. The result is a stronger concrete.

**5.7 Composites**   Composites are materials composed of two different materials bonded together in such a way that one serves as the matrix and surrounds fibres or particles of the other. Composites may take a number of forms:

1   Random particles in a matrix (Figure 5.20(a)).
2   Short, discontinuous fibres all lined up in the same direction in a matrix (Figure 5.20(b)).
3   Short, discontinuous fibres randomly orientated in a matrix (Figure 5.20(c)).
4   Long, continuous, fibres all lines up in the same direction in a matrix (Figure 5.20(d)).

Forms 1 and 2 give composites with mechanical properties the same in all directions, while 3 and 4 give composites with properties in the direction of the fibres different from those at right angles to them. In addition we need to include laminates made of sandwiched materials.

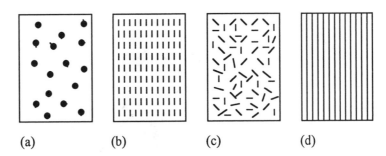

(a)          (b)          (c)          (d)

Figure 5.20 *Examples of composite structures*

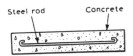

Figure 5.21 *Reinforced concrete*

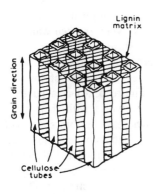

Figure 5.22 *Wood*

There are many examples of composite materials encountered in everyday products. For example, there are composites involving glass fibres, carbon fibres and glass beads in polymers, ceramic fibres in metals and metal fibres in ceramics. Many plastics are glass fibre or glass particle reinforced, the result being a much stronger and stiffer material than given by the plastic alone. Polymer composites are used for such applications as car instrument panels and lamp housings, domestic shower units and crash helmets. Vehicle tyres are rubber reinforced with woven cords. A common example of a metal reinforced ceramic composite is reinforced concrete. This has steel rods embedded in concrete (Figure 5.21). The composite material enables loads to be carried that otherwise could not have been carried by the concrete alone. Ceramic and metal composites are used for rocket nozzles, wire-drawing dies, cutting tools and other applications where hardness and performance at high temperatures might be required. Cermets, widely used for the tips of cutting tools, are composites involving ceramic particles in a metal matrix. Ceramic particles are hard but brittle and lack toughness, the metal is soft and ductile. Embedding the ceramic particles in the metal gives a material that is strong, hard and tough.

There are many naturally occurring composites. For example, wood is a natural composite material with tubes of cellulose bonded by a natural plastic called lignin (Figure 5.22). Natural fibres, such as cotton, wool and silk, are other examples of naturally occurring composites. The fibres consist of cellulose fibres in a matrix.

Foams can be regarded as a form of particulate composite in which the component bound by the matrix is not a solid but bubbles of gas. Such foams are used as cushioning in furniture, energy-absorbent packaging, thermal insulation and for buoyancy. The properties of foams are determined by the cellular structure and the ratio of the bulk density of the foam to that of the unfoamed matrix. The cellular structure can be open cell, closed cell, or a mixture of the two. With a closed-cell structure, the gas bubbles in the foam are discrete and not interconnected, whereas with an open-cell structure the bubbles have coalesced and are interconnected.

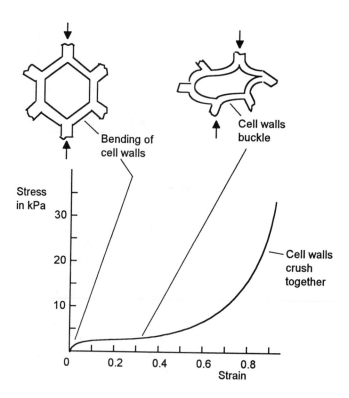

Figure 5.23  *Compressive stress–strain graph for a foam*

Figure 5.23 shows a typical compressive stress–strain graph for a foam. At low stresses the cell walls just bend. A marked change occurs in the stress–strain graph when the stress becomes high enough for the cell walls to buckle. The foam squashes a lot for very little change in stress. This buckling is elastic with the foam springing back to its original shape when the stress is removed. Eventually a point is reached when the cell walls are crushed together and the foam then becomes much stiffer.

Composites are generally regarded as fibre- or particle-reinforced materials. However, laminates are also composites. Examples of this are plywood and corrugated cardboard. Plywood (Figure 5.24) is a lamellar composite made by gluing together thin sheets of wood with their grain directions at right angles to each other. Thus, whereas the thin sheet had properties that were directional, the resulting laminate has no such directionality. Corrugated cardboard is another form of laminated structure (Figure 5.25), consisting of paper corrugations sandwiched between layers of paper. The resulting structure is much stiffer, in the direction parallel to the corrugations, than the paper alone.

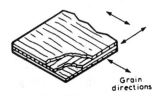

Figure 5.24  *Plywood*

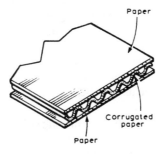

Figure 5.25 *Corrugated cardboard*

### 5.7.1 Fibres in a matrix

The fibres used in a matrix can be continuous long lengths all aligned parallel to an axis of the material, like the steels rods in reinforced concrete, or short fibres randomly orientated in the material. The long length fibres give a directionality to the properties, the tensile strength and tensile modulus being much higher along the direction of the fibres than at right angles. Randomly orientated short fibres do not lead to this directionality of properties but do not offer such high tensile strengths or tensile modulus values. For example, a glass fibre reinforced plastic (polyester) with long fibres might have a tensile strength of 800 MPa in the direction of the fibres and only 30 MPa at right angles to them. With short fibres the tensile strength in all directions might be 110 MPa.

Consider a composite rod made up of continuous fibres, all parallel to the rod axis, in a matrix (Figure 5.26). When tensile forces are applied to the composite rod, then each element in the composite has a share of the applied forces. Thus:

total force = forces on fibres + force on matrix

But since stress = force/area, then the force on the fibres is equal to the product of the stress $\sigma_f$ on the fibres and their total cross-sectional area $A_f$. Likewise, the force on the matrix is equal to the product of the stress $\sigma_m$ on the matrix and its cross-sectional area $A_m$. Hence:

total force = $\sigma_f A_f + \sigma_m A_m$

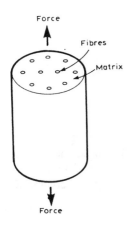

Figure 5.26 *Continuous fibres in a matrix*

Dividing both sides of the equation by the total area $A$ of the composite gives:

$$\text{stress on composite} = \frac{\text{total force}}{\text{total area}} = \sigma_f \frac{A_f}{A} + \sigma_m \frac{A_m}{A}$$

Thus the stress on the composite is the stress on the fibres multiplied by the fraction of the area that is fibres plus the stress on the matrix multiplied by the fraction of the area that is matrix.

Suppose we have glass fibres with a tensile strength of 1500 MPa in a matrix of polyester with a tensile strength of 45 MPa. If the fibres occupy, say, 60% of the cross-sectional area of the composite, then the above equation indicates that the tensile strength of the composite, i.e. the stress the composite can withstand when both the fibres and matrix are stressed to their limits, will be:

$$\text{strength of composite} = 1500 \times 0.6 + 45 \times 0.4 = 918 \text{ MPa}$$

The composite has a much higher tensile strength than that of the polyester alone.

If the fibres are firmly bonded to the matrix, then the elongation of the fibres and matrix must be the same and equal to that of the composite as a whole. Thus:

$$\text{strain on composite} = \text{strain on fibres} = \text{strain on matrix}$$

Dividing the stress equation above by this strain gives, since stress/strain is the tensile modulus,

$$\text{modulus of composite} = E_f \frac{A_f}{A} + E_m \frac{A_m}{A}$$

Suppose we have glass fibres with a tensile modulus of 76 GPa in a matrix of polyester having a tensile modulus of 3 GPa. If the fibres occupy, say, 60% of the cross-sectional area of the composite then the tensile modulus of the composite is:

$$\text{modulus of composite} = 76 \times 0.6 + 3 \times 0.4 = 46.8 \text{ GPa}$$

The composite has a tensile modulus much higher than that of the polyester alone.

### Example

A column of reinforced concrete has steel reinforcing rods running through the entire length of the column and parallel to the axis of the column. If the concrete has a modulus of elasticity of 20 GPa and the steel 210 GPa, what is the modulus of elasticity of the column if the steel rods occupy 10% of the cross-sectional area?

$$\text{Modulus of composite} = E_f \frac{A_f}{A} + E_m \frac{A_m}{A}$$

$$= 210 \times 0.1 + 20 \times 0.9 = 39 \text{ GPa}$$

**Example**

Carbon fibres with a tensile modulus of 400 GPa are used to reinforce aluminium with a tensile modulus of 70 GPa. If the fibres are long and parallel to the axis along which the load is applied, what is the tensile modulus of the composite when the fibres occupy 50% of the composite area?

$$\text{Modulus of composite} = E_f \frac{A_f}{A} + E_m \frac{A_m}{A}$$

$$= 400 \times 0.5 + 70 \times 0.5 = 235 \text{ GPa}$$

### 5.7.2 Manufacturing fibre-reinforced plastic composites

The following are some of the methods that are used industrially for the manufacture of fibre-reinforced plastic composites:

1  *Hand lay-up*
   The fibres might be in the form of chopped-strand mats or woven fabrics. These are placed in a mould shaped to the form required of the finished product (Figure 5.27(a)). A liquid thermosetting resin is then mixed with a curing agent and applied with a brush or roller to the fabric. Layers of fabric impregnated with the resin are used to build up the required thickness. Curing, i.e. waiting for the thermosetting polymer bonds to form a network, is usually at room temperature. Such a method is particularly suited to one-offs or small production runs of such items as the hulls of boats.

2  *Spray-up*
   This method involves chopped fibres, resin and hardener being sprayed on to a mould. To remove trapped air, the sprayed composite has to be rolled before the resin cures.

3  *Sheet moulding*
   Layers of fibres are pre-impregnated with resin and partially cured, such a material being referred to as sheet moulding compound (SMC). These sheets have a shelf-life of some three to six months at room temperature. The sheets are stacked on the open mould surface and then forced into the mould and the required shape before being fully cured. Figure 5.27(b) illustrates this when a pair of matched dies are used. This forcing can also be done by vacuum forming, the air between a sheet and a mould surface being removed and so the atmospheric pressure on the other sides of the sheet forces it against the mould surface.

4 *Dough moulding*

This method involves using dough moulding compound (DMC). This is a blend of short fibres and resin that has the consistency of bread dough or putty. The DMC is pressed into a open mould and then cured.

5 *Resin transfer moulding*

This involves fibres and resin being mixed and then injected under pressure into a closed mould before being cured (Figure 5.27(c)). This method is used for such products as fan blades, water tanks, seating, bus shelters and machine cabinets.

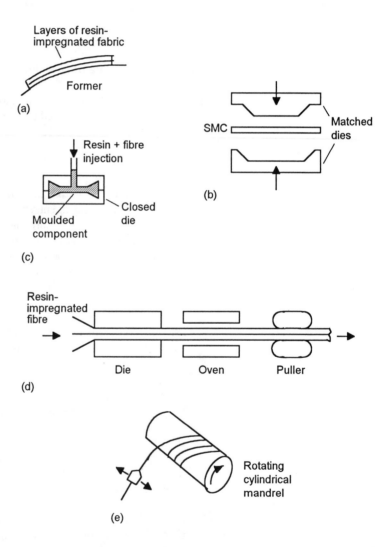

Figure 5.27 *(a) Hand lay-up, (b) matched die moulding, (c) resin transfer moulding, (d) pultrusion process, (e) filament winding*

6 *Pultrusion*

This method is used for the production of long lengths of uniform cross-section rods, tubes or I-sections. Continuous lengths of fibre are passed through a bath of resin and then pulled through a heated die to give the required shape product (Figure 5.27(d)).

7 *Filament winding*

Continuous lengths of fibres are passed through resin and laid out in the required directions on a mandrel (Figure 5.27(e)). This is generally done using a computer-controlled system. After curing, the mandrel is removed. Products made using this method are pressure vessels, helicopter blades and storage tanks.

## 5.8 Semiconductors

Electrical conductors, i.e. metals, have high electrical conductivity, whereas insulators have very low conductivity. Semiconductors fall between these two extremes. A brief discussion of electrical conductivity in these materials was given in Section 4.5. Here we will be concentrating on how the electrical conductivity of semiconductors can be modified by doping.

A simple theoretical model to explain conductivity and how it is affected by doping is called *band theory*. The theory identifies the energy levels available for electrons in solids. In the model of an atom as consisting of electrons in orbit about a positively charged nucleus, the electrons can only exist in particular orbits and not at just any distance from the nucleus. Each orbit corresponds to a specific level of energy. Thus the electrons can only exist at particular energy levels. In the case of a solid, when atoms are packed close together, these levels merge to give bands of energy. Electrons involved in the valence bonding occupy energy levels in the *valence band*. The valence electrons are still tied by some form of bonding to atoms. Electrons that have gained enough energy to be released from being tied to atoms are said to be in the *conduction band*. The differences between good conductors, insulators and semiconductors can be explained in terms of the size of the energy gap between the valence and conduction bands and hence the energy the valence electrons need to be able to participate in conduction.

In the case of a metal, a good conductor, we have the valence electrons in an energy band which overlaps with a conduction energy band (Figure 5.28). Valence electrons are thus able to accept very small amounts of energy and move into vacant higher energy levels. Typically, a metal will have about $10^{28}$ atoms per cubic metre and each atom will have one valence electron which is released for conduction. Thus there are about $10^{28}$ electrons per cubic metre able to move in the conduction band. When a potential difference is applied across a length of metal wire an electric field is produced in the wire and causes the electrons to move and so give a current. Think of the electrons as being rather like a small number of people in a large square. Given energy they can also move easily.

With an insulator we have, at a temperature of absolute zero, the valence electrons in a full energy band that does not overlap with the conduction

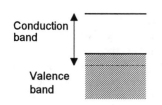

Conduction band

Valence band

Figure 5.28 *Energy bands for a metal*

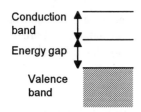

Figure 5.29 *Energy bands for an insulator*

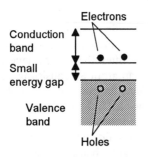

Figure 5.30 *Energy bands for a semiconductor*

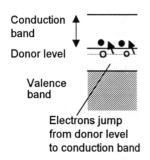

Figure 5.31 *n-type semiconductor*

band and has a large energy gap between them (Figure 5.29). The gap is too large for any electrons at room temperature to have sufficient energy to have jumped from the valence band into the conduction band so there are no electrons in the conduction band. For electrons to move through the material and give electrical conduction, they must be able to move into vacant energy levels. Think of this as people packed together on one bank of a river. Before they can accept energy and move, they have to have enough energy to jump across the river to the empty bank opposite. With metals there is no gap between the valence and conduction bands, but with insulators the energy gap is high and too big for thermal energy at room temperature to get electrons across it. Thus at room temperature, an insulator has a very low conductivity.

With a pure semiconductor, we have a conduction band only a small distance above the valence band (Figure 5.30). At room temperature, some of the valence electrons have received sufficient energy to jump across the gap into the conduction band. We then have a few electrons in the conduction band that are capable of accepting small amounts of energy and so participating in the conduction process. But we also have holes in the valence band, i.e. the sites from which valence electrons jumped to the conduction band. This means that, when an electric field is applied, some movement of electrons can also occur in the valence band. We can think of the holes moving as an electron jumps into a hole, leaving another hole behind into which another electron can jump, and so on. There will be the same number of electrons in the conduction band as holes in the valence band. Such a semiconductor is said to be *intrinsic*. With a pure semiconductor, we can thus ascribe a current as being due in equal parts to electron movement and hole movement. In silicon, a semiconductor, there are at a temperature of 300 K some $1.4 \times 10^{16}$ electrons per cubic metre in the conduction band and $1.4 \times 10^{16}$ holes per cubic metre in the valence band.

### 5.8.1 Doping

The balance between the number of electrons and holes can be changed by replacing some of the semiconductor atoms in the solid by atoms from other elements. This process is known as *doping* and typically about one atom in ten million might be replaced.

The silicon atom has four electrons which participate in bonding (see, for example, Figure 4.23). If atoms of an element having five electrons for participation in bonding are introduced into silicon, then there is a spare electron since only four electrons can participate in bonds within the silicon lattice. This electron is thus easily made available for conduction. Elements which donate electrons in this way are termed *donors*. In terms of the energy band model, the extra electron is in an energy level that falls in the energy gap between the valence and conduction bands of silicon but very close to the conduction band (Figure 5.31). This energy gap between donor level and conduction band is so small that at room temperature virtually all the donor electrons have moved into the conduction band. Thus there are more electrons in the conduction band than holes in the valence band and so electrical conduction with such a doped material is more by electrons than holes. For this reason, this form of doped semiconductor is called *n-type*, the n indicating that the conduction is predominantly by negative charge carriers. Arsenic, antimony and phosphorus are examples of elements that are added to silicon to give n-type semiconductors.

If an element having atoms with just three electrons which participate in bonding is added to silicon, then all three of its electrons participate in bonds with silicon atoms. However, there is a deficiency of one electron and so one bond with a silicon atom is incomplete. A hole has been introduced. Elements that supply holes in this way are termed *acceptors*. With the energy band model, an energy level is introduced into the gap between the valence and conduction bands of the silicon but very close to the valence level (Figure 5.32). This gap between the acceptor energy level and the valence band is so small that at room temperature it will have been filled by the movement of electrons from the valence band. There are thus more holes introduced into the valence band so that we have more holes in the valence band than electrons in the conduction band. Electrical conduction with such a doped material is more by holes than electrons. For this reason, this form of doped semiconductor is called *p-type*, the p being because conduction is predominantly by positive charge carriers. A hole is considered to behave like a positive charge carrier in that when an electric field is applied to a material, a hole moves in the opposite direction to an electron. Boron, aluminium, indium and gallium are examples of elements that are added to silicon to give p-type semiconductors.

Another way of picturing this is shown in Figure 5.33. The lines joining the atoms represent the electrons shared in bonding and so, at absolute zero, tightly bound to atoms. Silicon has four outer electrons involved in such bonding. When phosphorus is added, a spare electron is introduced. Phosphorus-doped silicon is n-type. When aluminium is added, a spare hole is introduced. Aluminium-doped silicon is p-type.

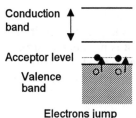

Figure 5.32 *p-type semiconductor*

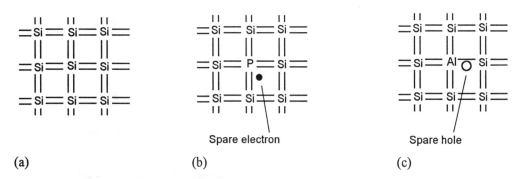

(a)   (b)   (c)

Figure 5.33 *(a) Pure silicon, (b) silicon doped with phosphorus, (c) silicon doped with aluminium*

Such doped semiconductors, n-type and p-type, are termed *extrinsic* meaning that there will be a majority charge carrier and a minority charge carrier. With n-type material, the majority charge carrier is electrons in the conduction band and for p-type it is holes in the valence band. Typically, doping replaces about one in every ten million atoms, i.e. 1 in $10^7$. Since there are about $10^{28}$ atoms per cubic metre, about $10^{21}$ dopant atoms per cubic metre will be used. Each dopant atom will donate one electron or provided on hole. Thus there will be about $10^{21}$ electrons donated or holes provided per cubic metre. Since intrinsic silicon has about $10^{16}$ conduction electrons and holes per cubic metre, the doping introduces considerably more charge carriers and swamps the intrinsic charge carriers. The majority charge carriers are thus considerably in excess of the minority charge carriers.

Thus when a potential difference is connected across a piece of semi-conductor, if it is n-type then the resulting current is largely the result of the movement of electrons in the conduction band. If it is p-type, the current is largely the result of movement of holes in the valence band. In terms of the picture presented in Figure 5.33, we can imagine the situation to be as shown in Figure 5.34.

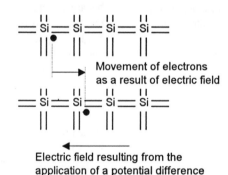

(a)

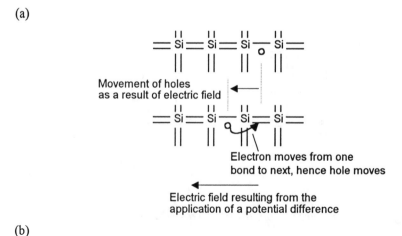

(b)

Figure 5.34  *Current with (a) n-type, (b) p-type semiconductors*

### 5.8.2 pn junction diode

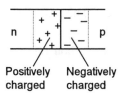

n-type        p-type

Electron movement

Hole movement

Figure 5.35 *p-type and n-type semiconductors brought into contact*

n    p

Positively     Negatively
charged        charged

Figure 5.36 *The pn junction*

The model used above can be used to give a simple explanation of the pn junction diode. Consider what happens if we have a n-type semiconductor in contact with a p-type semiconductor. Before contact we have two materials that are electrically neutral, i.e. in each the amount of positive charge equals the amount of negative charge. However, in the n-type semiconductor we have electrons available for conduction in the conduction band and in the p-type material we have holes available for conduction in the valence band. When the two materials are in contact then electrons from the n-type semiconductor can diffuse across the junction and into holes in the p-type semiconductor (Figure 5.35). We can also consider the holes to be diffusing across the junction in the opposite direction. Because electrons leave the n-type semiconductor it is loosing negative charge and so ends up with a net positive charge. Because the p-type material is gaining electrons it becomes negatively charged. Thus electrons and holes diffuse across the junction until the build-up of charge on each material is such as to prevent further charge movement. The result is shown in Figure 5.36. This separation of charge in the vicinity of the junction gives a result which is similar to a battery. When an external potential difference is connected across a pn junction, we can easily get electrons to flow through the circuit when they flow in the direction from p to n, in the reverse direction they are opposed by the charges at the junction. Thus we have a device that allows current flow in just one direction.

## Problems

*Questions 1 to 12 have four answer options: A, B, C and D. Choose the correct answer from the answer options.*

1  Increasing the size of the grains in an alloy:

A  Increases the tensile strength.
B  Increases the tensile modulus.
C  Increases the ductility.
D  Increases the hardness.

2  Decide whether each of these statements is TRUE (T) or FALSE (F).

Mild steel is more ductile than medium-carbon steel because:
(i)  It is predominantly a ferrite structure.
(ii)  It has a smaller percentage of carbon.

A  (i) T  (ii) T
B  (i) T  (ii) F
C  (i) F  (ii) T
D  (i) F  (ii) F

**3**   Decide whether each of these statements is TRUE (T) or FALSE (F).

A treatment of a metal that results in the density of dislocations increasing would be expected to:
(i)   Increase the brittleness of the metal.
(ii)  Increase the strength of the metal.

A   (i) T   (ii) T
B   (i) T   (ii) F
C   (i) F   (ii) T
D   (i) F   (ii) F

**4**   The effect of the annealing heat treatment for an alloy is to:

A   Increase the tensile strength.
B   Increase the tensile modulus.
C   Increase the ductility.
D   Increase the hardness.

**5**   Decide whether each of these statements is TRUE (T) or FALSE (F).

The stiffness of a polymer can be increased by:
(i)   Incorporating molecular blocks, such as a six-carbon ring, in the carbon backbone to the chain.
(ii)  Replacing some of the hydrogen atoms attached to the carbon backbone by bulky atoms, such as chlorine.

A   (i) T   (ii) T
B   (i) T   (ii) F
C   (i) F   (ii) T
D   (i) F   (ii) F

**6**   Decide whether each of these statements is TRUE (T) or FALSE (F).

Unfolding the linear polymer chains of a crystalline polymer so that the chains are all orientated in the same direction gives a material which is:
(i)   Weaker.
(ii)  Less stiff.

A   (i) T   (ii) T
B   (i) T   (ii) F
C   (i) F   (ii) T
D   (i) F   (ii) F

**7**   Decide whether each of these statements is TRUE (T) or FALSE (F).

In a fibre-reinforced polymer composite, the matrix has:
(i)   A higher tensile modulus than the fibres.
(ii)  A higher tensile strength than the fibres.

A   (i) T   (ii) T
B   (i) T   (ii) F
C   (i) F   (ii) T
D   (i) F   (ii) F

**8** Decide whether each of these statements is TRUE (T) or FALSE (F).

A composite with a polymer matrix reinforced with long fibres all orientated in the same direction will have:
(i) A tensile modulus that differs in a direction at right angles to the fibres from that in the direction of the fibres.
(ii) A tensile strength that differs in a direction at right angles to the fibres from that in the direction of the fibres.

A  (i) T  (ii) T
B  (i) T  (ii) F
C  (i) F  (ii) T
D  (i) F  (ii) F

**9** Decide whether each of these statements is TRUE (T) or FALSE (F).

Ceramic matrix composites reinforced with metals have:
(i) A lower toughness than that of the ceramic alone.
(ii) A lower tensile strength than that of the ceramic alone.

A  (i) T  (ii) T
B  (i) T  (ii) F
C  (i) F  (ii) T
D  (i) F  (ii) F

**10** Decide whether each of these statements is TRUE (T) or FALSE (F).

An intrinsic semiconductor at room temperature has:
(i) The same number of holes in the valence band as electrons in the conduction band.
(ii) Electrical conduction primarily as a result of electron movement in the conduction band.

A  (i) T  (ii) T
B  (i) T  (ii) F
C  (i) F  (ii) T
D  (i) F  (ii) F

**11** Decide whether each of these statements is TRUE (T) or FALSE (F).

A p-type semiconductor has:
(i) More holes in the valence band than electrons in the conduction band.
(ii) An energy level supplied by the donor atoms which is between the valence and conduction bands but closer to the valence band.

A  (i) T  (ii) T
B  (i) T  (ii) F
C  (i) F  (ii) T
D  (i) F  (ii) F

12 Decide whether each of these statements is TRUE (T) or FALSE (F).

Doping silicon with a material having three valence electrons results in:
(i) A n-type semiconductor.
(ii) An energy level being introduced between the valence and conduction bands.

A  (i) T  (ii) T
B  (i) T  (ii) F
C  (i) F  (ii) T
D  (i) F  (ii) F

13 Explain what is meant by the term *alloy*.

14 Explain the terms *ferrous alloy* and *non-ferrous alloy*.

15 How does the grain size in a metal affect its properties?

16 How does the shape of grains within a metal affect its properties?

17 Describe the effects on the grain structure and properties of a metal of cold working.

18 Describe the effects on the properties of carbon steels of increasing the percentage of carbon on the alloy.

19 What types of structure might you expect for a metal that is (a) ductile, (b) brittle?

20 A pure metal is formed into an alloy by larger atoms being forced into the spaces in its crystal structure. What changes might be expected in the properties and why?

21 Describe how the mechanical properties of a cold-worked metal changes as its temperature is raised from room temperature to about $0.6T_m$, where $T_m$ is the melting point on the kelvin scale.

22 How does the temperature at which working is carried out determine the grain size and so the mechanical properties?

23 Describe the difference between amorphous and crystalline polymer structures and explain how the amount of crystallinity affects the mechanical properties of the polymer.

24 Why are the following added to polymers: (a) stabilisers, (b) plasticisers and (c) fillers?

25 Increasing the amount of sulphur in a rubber increases the amount of cross-linking between the molecular chains. How does this change the properties of the rubber?

**26** Explain how elastomers can be stretched to several times their length and still be elastic and return to their original length.

**27** Calculate the tensile modulus of a composite consisting of 45% by volume of long aligned glass fibres, tensile modulus 76 GPa, in a polyester matrix, tensile modulus 4 GPa. In what direction does you answer give the modulus?

**28** In place of the glass fibres referred to in problem 27, carbon fibres are used. What would be the tensile modulus of the composite if the carbon fibres had a tensile modulus of 400 GPa?

**29** Long boron fibres, tensile modulus 340 GPa, are used to make a composite with aluminium as the matrix, the aluminium having a tensile modulus of 70 GPa. What would be the tensile modulus of the composite in the direction of the aligned fibres if they constitute 50% of the volume of the composite?

**30** How will the properties of composites differ if they are (a) made of long fibres all orientated in the same direction, (b) short fibres with random orientation?

# 6 Materials and their uses

## 6.1 Materials and their uses

In this chapter the uses made of materials are considered and which materials would be suitable for particular purposes. Materials have evolved over the years and the materials we now have available offer opportunities for the fashioning of products of a quality, and at a cost, that were not available in years gone by.

### 6.1.1 The evolution of materials

The early history of the human race can be divided into periods according to the materials that were predominantly used. Thus we have the Stone Age, the Bronze Age and Iron Age.

In the *Stone Age* (about 8000 BC to 4000 BC), people could only use the materials they found around them such as stone, wood, clay, animal hides, bone, etc. The tools they made were limited to what they could fashion out of these materials. Thus tools were limited to those that could be made from stone, flint, bone and horn.

By about 4000 BC, people in the Middle East were able to extract *copper* from its ore and it rapidly became an important material. Copper is a ductile material which can be hammered into shapes, thus enabling a greater variety of items to be fashioned than was possible with stone. Because the copper ores contained impurities that were not completely removed by the smelting, alloys were produced. It was soon realised that the deliberate adding of additives to copper could produce materials with improved properties. About 2000 BC it was found that when tin was added to copper, an alloy was produced that had an attractive colour, was easy to form and harder than copper alone. This alloy was called *bronze*. Thus we have the *Bronze Age*.

About 1200 BC the extraction of *iron* from its ores signalled another major development, hence the *Iron Age*. Iron in its pure form was, however, inferior to bronze but by heating items fashioned from iron in charcoal and hammering them, a tougher material, called *steel*, was produced. Plunging the hot metal into cold water, i.e. quenching, was found to improve the hardness. It was also found that reheating and cooling the metal slowly produced a less hard but tougher and less brittle material, this process now being termed tempering. Thus *heat treatment processes* were developed.

The large-scale production of iron can be considered an important development in the evolution of materials in that it made the material more widely and cheaply available for products. Large-scale iron production with the first coke-fuelled blast furnace started in 1709. Cast iron was used in 1777 to build a bridge at the place in England now known as Ironbridge.

The term *industrial revolution* is used for the period that followed as the pace of developments of materials and machines increased rapidly and resulted in major changes in the industrial environment and the products generally available. The year1860 saw the development of the Bessemer and open hearth processes for the production of steel, and this date may be considered to mark the general use of steel as a constructional material. Aluminium was extracted from its ores in 1845 and produced commercially in 1886. In the years that followed, many new alloys were developed. The high strength aluminium alloy Duralumin was developed in 1909, stainless steel in 1913, high strength nickel–chromium alloys for high temperature use in 1931. Titanium was first produced commercially in 1948.

While naturally occurring *plastics* have been used for many years, the first manufactured plastic, celluloid, was not developed until 1862. In 1906 Bakelite was developed. The period after about 1930, often termed the *Plastic Age*, saw a major development of plastics and their use in a wide range of products. Polyethylene is an example of a scientific investigation yielding a surprise result. In 1931 a Dutch scientist, A. Michels, was given approval by ICI to design apparatus that could be used to carry out research into the effects of high pressure on chemical reactions. In 1932 the investigation began and in March 1933 a surprise result was obtained. The chemical reaction between ethylene and benzaldehyde was being studied at a pressure of 2000 times atmospheric pressure and a temperature of 170°C when a waxy solid was found to form. The material that had been formed was *polyethylene*. The experiment had not been designed to develop a new material but that was the outcome. The commercial production of polyethylene started in England in 1941. The development of *polyvinylchloride* was, however, an investigation where a new material was sought. In 1936 there was no readily available material that could replace natural rubber. In the event of a war Britain's natural rubber supply would be at risk since it had to travel by sea from the Far East. Thus a substitute was required and research was initiated. In July 1940 a small amount of PVC was produced, but there was to be many problems before commercial production of PVC with suitable properties could begin in 1945 .

The evolution of materials over the years has resulted in changes in our life-styles. Thus when tools were limited to those that could be fashioned out of stone, there was severe limitations on what could be achieved with them. The development of metals enabled finer products to be fashioned. For example, bronze swords were far superior weapons to stone weapons. Consider what the world would be like today if plastics had not been developed.

In 1930 the world production of steel was about 300 000 million kg, 2 000 million kg of copper and zinc and there was virtually no aluminium or plastics. By 1950 there had been only a slight increase in the production of steel, zinc and copper but aluminium has risen to some 1000 million kg and plastics to 1000 million kg. Another twenty years later, 1970, steel, zinc and copper, had showed only a small increase but aluminium had now risen to equal the production of copper and zinc, at about 10 000 million kg and plastics had overtaken them to become 20 000 million kg. By 1990 the amount of plastic used had increased even more.

## 6.2 Materials selection

What functions does a product have to perform? This is a question that requires an answer before possible materials can be considered. From this stems the conditions under which the materials will be used and the properties required of them.

Consider the problem of the material for a domestic kitchen pan. The functions of a domestic kitchen pan may be deemed to be: to hold liquid and allow it to be heated to temperatures of the order of 100°C. From a consideration of the functions, we can arrive at the basic design requirements. Thus, a consequence of these functions for the pan are the requirements for a particular shape of container that must not deform when heated to these temperatures. It must be a good conductor of heat. It must be leak-proof. It must not ignite when in contact with a flame or hot electrical element. In addition there may be other requirement that are not so essential, but certainly desirable. For the pan we might thus require an attractive finish for it.

From these requirements we can now define the required properties of the materials. Thus for the pan, the requirement that the material be a good conductor of heat would seem to reduce the consideration to metals, particularly when taken together with the requirement that the material can be put in contact with a flame and contain hot liquids. This would effectively rule out polymers. But what properties are required of the metal? For the body of the pan to be shaped as a single entity, a manufacturing process in which a sheet of the metal is pressed out into the required shape is suggested. This might be by a so-called deep drawing process (Figure 6.1) or by a similar process called pressing (see Figure 6.2, pressing is like drawing but with the edges of the material clamped). A ductile material is required in sheet form for such processes. Thus we might consider an aluminium alloy. Another possibility would be a stainless steel, stainless because rusty pans would not be very desirable. The deciding factor is likely to be cost, though there may be some prestige value attached to having stainless steel plans as opposed to aluminium, which would allow a higher price to be charged. For the same volume of material, the stainless steel will probably cost about three times the aluminium alloy.

The above represents one line of argument regarding the design of pans. It is instructive to examine a range of pans and consider the materials used and what reasons might be advanced for them being chosen. Why, for example, are some pans made of glass, of a ceramic, of a steel coated on the outside with an enamel and on the inside with a non-stick polymer polytetrafluoroethylene (PTFE)?

The above is only the consideration of the container part of the pan. There is still the handle to consider. The function required is that it can be used to lift the pan and contents, even when they are hot. The properties required are thus poor thermal conductivity, ability to withstand the temperatures of the hot pan, stiffness and adequate strength. The handle can be considered as a cantilever with a load, the pan and contents, at its free end. Before going too far with considering the design and materials for the handle, British standards can be consulted. BS 6743 gives a standard specification for the performance of handles and handle assemblies attached to cookware. This sets the levels of safe performance against identified tests

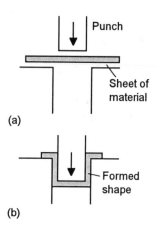

(a)

(b)

Figure 6.1 *Deep drawing*

simulating hazards experienced in normal service. The need for the handle to have low thermal conductivity indicates that metal would not be a good choice. A polymer is thus a possibility. It needs, however, to be able to withstand a temperature of the order of 100°C at the pan end, have a reasonably high modulus of elasticity and reasonable strength. These requirements suggest a thermoset is more likely to be feasible than a thermoplastic. A possibility is phenol formaldehyde (Bakelite). The dark colour of this material is no problem in these circumstances. When filled, with say wood flour, it has a high enough maximum service temperature of about 150°C, a tensile modulus of 5.0 to 8.0 GPa (high for a polymer), and a tensile strength of 40 to 55 MPa.

In the above considerations of the pan and the handle the item that has so far not been discussed is the life of the items. The purchaser of the pan wants it to last without problems for a reasonable period of time. This is likely to be years. The handle should not break during this time, discolour or deteriorate when used and washed a large number of times. The pan should not wear thin or change its mechanical properties with frequent heating, exposure to hot liquids and washing-up liquids.

### 6.2.1 Stages in the selection process

As the above examples indicate, there are a number of stages involved in arriving at possible materials and processing requirements for a product. These can be summarised, in very simple terms, as follows:

1   Define the functions required of the product.
2   Consider a tentative design, taking into account any codes of practice, national or international standards.
3   Define the properties required of the materials.
4   Identify possible materials, taking into account availability in the required forms.
5   Identify possible processes that would enable the design to be realised.
6   Consider the possible materials and possible processes and arrive at a proposal for both. If not feasible, reconsider the design and go back through the cycle.
7   Consider how the product will behave during its service life.

### 6.2.2 Costs

The total cost to the consumer of a manufactured article in service, i.e. the so-called *total life cost*, is made up of a number of items:

1   The purchase price. This includes the costs of production, the fixed costs arising from factory overheads, administration, etc., and the manufacturer's profit. The costs of production include the cost of the basic materials and the cost of manufacture.

2   The cost of ownership. This includes such costs as those associated with maintenance, repair and replacement.

## 6.3 Materials and products

The following are some case studies of materials selection for particular applications.

### 6.3.1 Car bodywork

The functions required of car bodywork are protecting the engine and car occupants from the weather and providing a pleasing appearance. The requirements for the material are thus that it can be formed to the shapes required, it has a smooth and shiny surface, corrosion is not too significant, it is sufficiently tough in service to withstand small knocks, it is stiff, and is cheap and can be mass produced.

The shapes required, together with the need for mass production, would suggest forming from sheet as the manufacturing process. This process involves pressing sheets over formers so that plastic deformation gives the required shape. Figure 6.2 shows the type of double-action press that is generally used. Hot forming does present the problem of an unacceptable surface finish and so a material has to be chosen that allows for cold forming. This means a highly ductile material. Possibilities would be low carbon steels or aluminium alloys, e.g. an alloy of aluminium with manganese. Low carbon steels are more ductile, though less strong, than high carbon steels (see Figure 5.2). In the annealed state, both these will give high ductility with percentage elongations of the order of 20 to 40%. On the basis of that information, both the steels and the aluminium alloys could be used. In addition, both are reasonably tough and when given a coat of paint are reasonably resistant to corrosion and so can be expected to have a reasonable length life.

Aluminium alloys have the advantage of lower densities and so could lead to lower weight cars. The carbon steel does, however, have some advantages, it work-hardens more than the aluminium alloy and so gives a harder material. In addition, the steel has a higher tensile modulus than the aluminium and so is stiffer. The great advantage of carbon steel, outweighing all other considerations, is that it is much cheaper than aluminium alloy. Typically it is about half the price. Thus the material generally used for car bodywork is a low carbon steel.

Polymer materials could be used for the car bodywork. The problem with such materials is obtaining enough stiffness – polymeric materials have tensile modulus values considerably smaller than metals. One way of overcoming this is to form a composite material with glass fibre or cloth in a matrix of a thermoset. This could be done using the hand lay-up process (see Section 5.7.2). Unfortunately, such a process of building up bodywork is a manual rather than machine process and so very slow and labour-intensive. While it can be used for one-off bodies, it is not suitable for mass production.

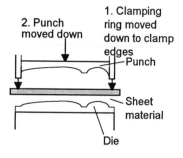

1. Clamping ring moved down to clamp edges

2. Punch moved down

Figure 6.2 *Forming sheets to the shape of the die by pressing*

### 6.3.2 Tennis rackets

The function of a tennis racket is to transmit power from the arm of the player to a tennis ball. The requirements for the frame and handle of a racket are a high strength, high stiffness, low weight, tough and able to withstand impact loading, durable and does not creep or warp as a result of exposure to temperature or humidity changes, and can be processed into the required shape. Another requirement that requires a little explanation is the ability to damp out vibrations. When the ball hits the strings, the impact leads to vibrations of the racket. These are then transmitted through the frame of the racket to the arm of the player. If these vibrations are not reduced in amplitude in this transmission, the elbow of the player can suffer some damage, known as tennis elbow. The elbow-joint does not like being vibrated. Cost will be a factor when considering tennis rackets for the general population but less a requirement for rackets for professional tennis players.

The requirement for high strength and low weight can be translated into a requirement for a high value of strength/density, i.e. specific strength. Similarly the requirement for high stiffness and low weight into a requirement for a high value of modulus/density, i.e. specific modulus. Possibilities would seem to be wood, metals and composites. Table 6.1 shows typical values for some possible materials.

Table 6.1 *Materials for tennis rackets*

| Material | Specific strength MPa/Mg m$^{-3}$ | Specific stiffness GPa/Mg m$^{-3}$ | Relative tougness | Relative vibration damping | Relative cost |
|---|---|---|---|---|---|
| *Woods* | | | | | |
| Ash | 107 | 20 | Good | Good | Low |
| Hickory | 105 | 21 | Good | Good | Low |
| *Aluminium alloys* | | | | | |
| Al–Cu alloy, precipitation hardened | 15 | 25 | Good | Poor | Medium |
| Al–Mg alloy, annealed | 54 | 25 | Good | Poor | Medium |
| *Steels* | | | | | |
| Mn steel, quenched and tempered | 90 | 27 | Good | Poor | Medium |
| Ni–Cr–Mo steel, quenched and tempered | 115 | 27 | Good | Poor | Medium |
| *Composites* | | | | | |
| Epoxy + 60% carbon | 890 | 90 | Medium | Medium | High |
| Epoxy + 70% glass | 750 | 25 | Medium | Medium | High |

Wood has the advantages that it is tough, has good specific strength, good damping properties for vibrations and is cheap. The specific stiffness could be better. Warping could be a problem. However, this can be overcome by using laminated wood, i.e. several pieces of wood with their fibres in different directions bonded together to give a laminate. This combining together of pieces of wood also gives a method by which the shape of the racket can be obtained.

Aluminium alloys have the advantages of toughness and good specific stiffness. They are, however, more expensive than wood. Another problem is that they have very poor vibration damping. Aluminium can be protected against corrosion attack by damp environments by anodising. An aluminium racket could be made by bending extruded hollow sections into the required shape.

Steels can give high specific strengths and high specific stiffness. The steels with these high strengths are likely to be comparable in price with the aluminium alloys. Problems are, however, the very poor vibration damping and the poor corrosion resistance in a damp environment. A steel racket could be made by bending extruded hollow sections into the required shape.

Composite materials can be made which have the advantages of very high specific strengths, very high specific stiffnesses, reasonable vibration damping and tolerable toughness. The major problem, however, is the high cost of such materials. A composite racket can be made by injecting a melt of a polymer containing carbon fibres into a racket-shaped mould. This would give a racket with a solid composite for the frame and handle. The procedure that can then be adopted to improve the properties is, while the racket is still in the mould and only the outer skin of the composite has solidified, to pour out the liquid core so that when the racket solidifies there is a hollow tube. The tube can then be filled with a polyurethane foam (Figure 6.3). This improves the vibration damping of the racket.

In comparing the above, the composite material racket gives the best properties but is considerably more expensive than the others. It thus is more likely to be used by the professional tennis player. For cheapness and properties, wood is probably the next best material, followed by aluminium alloys with steel being the worst.

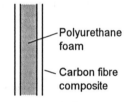

Figure 6.3  *Composite frame*

### 6.3.3 Small components for toys

Consider small components such as the wheels for, say, a small model toy car for use by a small child. The functions required of the wheels are that they are safe and rotate on their axles. The materials thus need to be non-toxic, reasonably tough, not easily deformed by knocks, not brittle and cheap. Before considering possible materials, there is a British standard, BS5665, which should be consulted. This specifies, in Part 1, material, construction and design requirements for toys, methods of test for certain properties, and requirements for packaging and marketing. Part 2 specifies categories of flammable materials not to be used in the manufacture of toys. Part 3 gives the requirements and methods of test for migration of

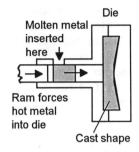

Figure 6.4 *Die casting*

antimony, arsenic, barium, cadmium, chromium, lead, mercury and selenium from toy materials.

The products are required to be cheap when produced in relatively large quantities and the products themselves are rather small. In the case of metals, a possible process is die casting. This process involves pouring or injecting liquid metal into a die, i.e. a metal mould (Figure 6.4). Though the initial cost of the die is high, a large number of  components can be produced from one die and so the cost per component becomes relatively low. In the case of polymers, a possible process is injection moulding. With this process, the polymer is melted and then forced into a mould. This also has a high die cost but large numbers of components can be produced from one die and hence the cost per component can be low. Both processes give a good surface finish and good dimensional accuracy.

In the case of metals, die casting limits the choice to those with relatively low melting points, i.e. aluminium, magnesium, zinc, lead and tin alloys. Table 6.2 shows relevant properties of these materials.

Table 6.2 *Die casting alloys at 20°C*

| Alloy | Density Mg/m$^3$ | Melting pt. °C | Strength MPa |
|-------|------------------|----------------|--------------|
| Aluminium | 2.7 | 600 | 150 |
| Lead | 11.3 | 320 | 20 |
| Magnesium | 1.8 | 520 | 150 |
| Tin | 7.3 | 230 | 12 |
| Zinc | 6.7 | 380 | 280 |

Safety considerations (see the British standard) rule lead out. Aluminium, magnesium and zinc are comparable in cost, with zinc tending to have the lower cost per unit weight. Tin is more expensive than these alloys. Zinc has a lower melting point than aluminium or magnesium and in the as-cast condition has the highest tensile strength. Thus zinc would seem to be the best metal choice for the product.

Zinc is very widely used for die casting. Its low melting point and excellent fluidity make it one of the easiest metals to cast. Small parts of complex shape and thin wall sections can be produced. Zinc alloys have relatively good mechanical properties and can be electroplated.

Polymers are a possible alternative to metals. The forming method capable of producing such items in quantity is injection moulding (Figure 6.5). With this method, a polymer is heated and then forced into a mould. The materials used are restricted to thermoplastics. The choice is then to polymers which are relatively stiff. Table 6.3 shows how the properties of possible polymers compare with those of zinc. The mechanical properties of the zinc alloy are superior to those of thermoplastics, it having higher strength, higher tensile modulus, and being tougher, more resistant to fatigue and creep. Where light weight is required, then polymers have the advantage, having densities of the order of one-sixth that of zinc. Where

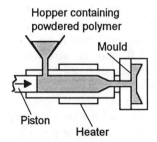

Figure 6.5 *Injection moulding*

coloured surfaces are required then polymers have the advantage since pigments can be incorporated in the polymer mix. However, if electroplating is required, then zinc has the advantage. On cost per unit weight then zinc is cheaper; however, the interest is likely to be cost per unit volume and on this basis polymers are likely to be cheaper.

On the basis of the above considerations, it is likely that zinc would be used where colour of surface is not a requirement but electroplating is required. Otherwise polymers, possibly ABS, would be used.

Table 6.3  *Comparison of zinc and thermoplastics at 20°C*

| Material | Strength MPa | Modulus GPa | Density Mg/m$^3$ | Relative cost/m$^3$ |
|---|---|---|---|---|
| ABS | 50 | 2.3 | 1.02–1.07 | 1 |
| Nylon 6 | 60 | 3.2 | 1.13–1.14 | 2 |
| Polycarbonate | 65 | 2.3 | 1.2 | 2 |
| Zinc alloy | 280 | 103 | 6.7 | 3 |

### 6.3.4 Electrical resistors

Consider the requirements for resistors for general use in electrical and electronic circuits. They are required to obey Ohm's law and have resistances that do not markedly change when the temperature changes. This would suggest metals as the required material; the resistance of semiconductors changes very markedly when the temperature changes. The values of resistance frequently required are in the range 1 kΩ to 10 MΩ. But metals tend to have resistivities of the order of $10^{-6}$ Ω m. Consider therefore the problem of using such a material for a resistance of 1 kΩ. Suppose we use wire with a diameter of 1 mm. The length $L$ of wire required will be given by:

$$\text{resistivity } \rho = 10^{-6} = \frac{RA}{L} = \frac{1000 \times \frac{1}{4}\pi \times 0.001^2}{L}$$

Hence the length required is 785 m. This is not practical. A more practical length, for a reasonably compact resistor, would be about 100 mm. This is only possible with such a resistivity if the cross-sectional area is $10^{-10}$ m$^2$. This would be a rectangular strip 0.01 mm by 0.01 mm or wire with a diameter of about 0.011 mm. How then can we produce such resistors?

One possible method by which we can use metals is to deposit very thin layers on an insulating substrate. This can then give a very thin thickness of metal. Thin films of thickness about 10 nm (1 nm = $10^{-9}$ m) are used. We can then make this into a reasonable length by etching a suitable pattern in the metal. Thus a spiral groove might be cut through the metal deposited on a cylindrical substrate (Figure 6.6). The resistance value can be adjusted to some required value by stopping the groove cutting when that value is

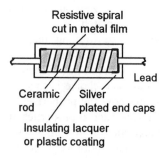

Resistive spiral
cut in metal film

Lead

Ceramic / Silver
rod      plated end caps

Insulating lacquer
or plastic coating

Figure 6.6  *Thin film resistor*

obtained. Nickel–chromium alloys (nichrome) are widely used for resistors manufactured in this way.

Another alternative is to mix a conductive powder with an insulator and organic solvent, the resulting mixture then being spread over an insulating substrate as a film about 10 μm thick (1 μm = $10^{-6}$ m). The conductive powders used are highly conductive oxides such as PdO or $RuO_2$. They have resistivities of about $10^{-6}$ Ω m and behave as metals. The mixture is fired so that organic solvents evaporate and the insulator and conductive material are left bonded to the substrate. The dispersed conductive particles are considered to form convoluted chains through the insulator. Thus we effectively end up with a number of exceedingly small cross-section conductors. The resistance value for the resistor is then determined by the concentration of the conductive powder in the insulator.

There is an alternative to using a metal and that is to use carbon in the form of graphite. Diamond is a crystal structure based solely on carbon atoms, each atoms being bound by covalent bonds to four other carbon atoms. The result is a strong three-dimensional structure that is an electrical insulator because there are no free electrons. Diamond is also very hard. Graphite, however, is a very soft material. It is the material used as the lead in pencils. Graphite, like diamond, consists only of carbon atoms. However, the way in which the carbon atoms are arranged in the solid is quite different (Figure 6.7). It can be considered to be a 'layered' structure. The atoms are strongly bonded together with covalent bonds in two-dimensional layers, with only very weak bonds, van der Waals bonds, between the atoms in different layers. As a result, there are free electrons between the layers and the material conducts electricity. At room temperature, the resistivity of diamond is about $10^{10}$ Ω m, while that of graphite is about $10^{-6}$ Ω m.

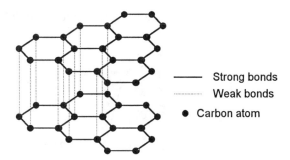

Figure 6.7  *Graphite*

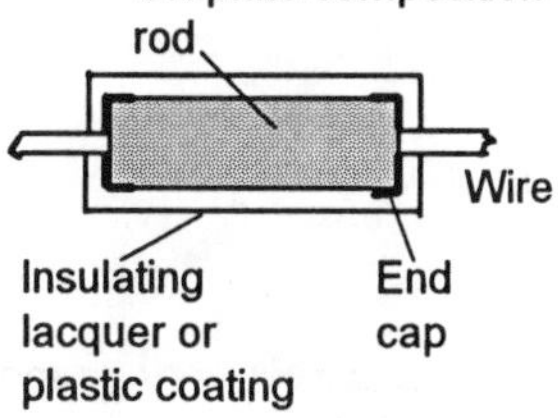

Resistors using graphite all tend to be about the same size, but can have a wide range of values. The process used for making carbon resistors involves powdered graphite being mixed with naphthalene and an insulating filler such as china clay. The mix is then pressed into little cylinders and then fired. This gets rid of most of the naphthalene and leaves a porous graphite structure. The amount of naphthalene, and hence the degree of porosity, determine the resistance of the cylinder. Figure 6.8 shows the form such a resistor might take.

Graphite composition rod

Wire

Insulating lacquer or plastic coating

End cap

Figure 6.8  *A graphite composition resistor*

### 6.3.5 Clothing fibres

The fibres used for clothing include cotton, wool and polymers such as nylon and terylene. What are the properties required of these materials which determine their behaviour when used in clothing? These properties might include high tensile strength, able to be stretched and recover without deformation, flexibility, ability to be dyed, reasonable weathering properties, resistant to shrinkage.

Figure 6.9 shows the types of stress–strain graph that are found with cotton, wool and nylon fibres. Wool stretches much more than nylon or cotton. Cotton has a higher tensile strength than nylon, this in turn having a higher strength than wool. Thus, of the three fibres, wool is the weakest and the one that stretches the most.

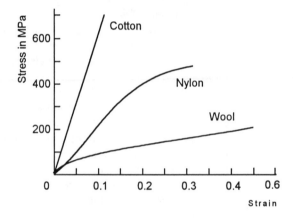

Figure 6.9  *Stress–strain graph for fibres*

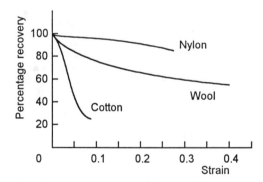

Figure 6.10  *Recovery graphs for fibres*

Linked with this stress–strain graph is the percentage recovery graph (Figure 6.10). A perfectly elastic material would have a 100% recovery, and spring back to its original length when the stress was removed. The recovery graph show how, for each of the materials, the percentage recovery depends on the strain acting on the material. Nylon is thus a more elastic material than wool, with wool being more elastic than cotton.

Thus if we compare wool with nylon, perhaps when used to make socks, the nylon socks would be stronger than the wool and the material more elastic. The wool socks would stretch more and be more readily permanently stretched than the nylon. On this basis then we might expect nylon socks to last longer and maintain their original shape longer.

Wool, cotton and nylon are all polymeric fibre materials, wool and cotton being natural fibres while nylon is a synthetic fibre. Synthetic polymer fibres, such as nylon, are made by melting pieces of the polymer and then forcing the liquid through orifices to give long lengths of extruded fibres (the process is rather like squeezing toothpaste out a toothpaste tube). The nylon fibres are then stretched to four or five times their initial length. This stretching orientates the polymer molecules along the length of the fibres and increases the strength, at the expense of the elongation possible.

### 6.3.6 Ceramic hobs

What are the properties required of the material used for a domestic ceramic hob? The functions of the hob is to be a decorative sheet of material under which heating elements raise 'hot spots', on which pans are placed, to high temperatures. The hot spots must be highly localised with no heat spreading out sideways. Thus a prime requirement is that the material used must be a poor thermal conductor. However, this conflicts with the requirement that the material at the hot spots conducts heat through from the heating elements to the pans. If there is not to be too great a temperature difference between the two sides of the sheet, we can only reasonably meet this requirement if the material is made thin. So the requirement seems to be for a thin sheet of a poor thermal conductor. This might suggest using a glass, since glass has a low thermal conductivity.

But there are other requirements. The hob must be able to withstand thermal shocks that might occur if cold water is poured on it when hot. Think what happens if you pour hot water into a cold glass. Because it is a relatively poor thermal conductor, the inside of the glass rises in temperature more rapidly than the outside. As a consequence, the inside of the glass expands more rapidly than the outside. This difference in expansion of the inside and outsides of the glass is very likely to result in the glass cracking. This effect can be minimised if we choose a glass with a very low coefficient of expansion. We can find such a glass. But there is still a problem with glass.

If you make a small nick in a glass rod it will break very readily at that point. Glass is not tough. Scratches, which inevitably would occur with a hob, would make a sheet of glass very susceptible to breaking because of a lack of toughness. The solution is a crystalline ceramic based on lithium aluminium silicate. This has a low thermal conductivity, a very low coefficient of expansion, and is reasonably tough. But there is still a problem to be overcome. Ceramics are not easy to fabricate in the required shape. This is overcome in this instance by making the hob as a glass, then using heat treatment to convert it into the crystalline form. Figure 6.11 illustrates the process.

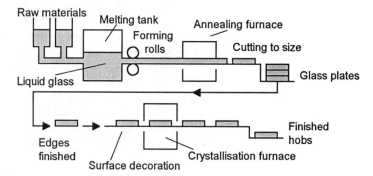

Figure 6.11  *Making ceramic hobs*

### 6.3.7 Plastic bottles

The properties required of the materials used for a bottle to contain, say, a fizzy drink might be (see Section 1.2):

1   It must have high impact strength, not be brittle and be tough.
2   The drink must not seep out through the container walls or loose its fizz due to the escape of carbon dioxide through the container walls.
3   The material must not taint the drink.
4   It must be relatively stiff.
5   It must be transparent and clear.
6   It must have a low density so that the bottle is light.
7   The material must be capable of being processed by blow moulding. Blow moulding is a process used by glass blowers to form bottles. A blob of red hot glass is formed on the end of a blowpipe and then air blown down the pipe to gives a hollow glass shape. A similar process is used with thermoplastics to form plastic bottles, the bottle being blown inside a hollow mould (Figure 6.12).

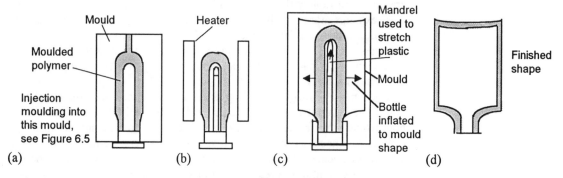

Figure 6.12  *(a) Injection moulding used to create basic shape, (b) the polymer is then heated to facilitate the stretching that occurs in (c), (d) the resulting bottle shape*

8   The material must be cheap since the bottle is designed to be thrown away after use.

The blow moulding requirements mean that the material must be a thermoplastic. Of the polymeric materials that are clear and transparent and reasonable tough, we are limited to low-density polyethylene, PVC or polyethylene terephthalate. Low-density polyethylene is the toughest. The material is required to have a low permeability to both water and carbon dioxide. All three have low permeability to water. Low-density polyethylene is, however, highly permeable to carbon dioxide and is ruled out on that count. PVC and polyethylene terephthalate have low permeabilities for carbon dioxide, with PVC being more permeable than polyethylene terephthalate. Polyethylene terephthalate is thus the choice for fizzy drinks containers. PVC is, however, widely used for non-carbonated drinks such as wine.

The bottle is designed to be thrown away after use. If recycling had been a requirement then a strong argument could have been developed for the use of glass bottles. There is very little recycling of plastics but a very significant amount of recycling of glass. Compared with plastics, the main disadvantages of glass are the higher weight of a glass bottle and the lower toughness.

### 6.3.8 Dental fillings

The prime function of dental fillings is to plug cavities in teeth. However, to this we might add that the result should as close as possible result in teeth in which the fillings cannot be perceived.

The properties required of the material used for dental fillings are thus:

1   Can be worked into small, drilled-out cavities in teeth.
2   After working into cavities it must set hard.
3   The hardened material must be abrasion-resistant, non-toxic and not deteriorate as a result of contact with saliva.
4   It should match the colour of the teeth.

The standard material that has been used for many years is an amalgam. The amalgam is made by mixing mercury with a silver–tin alloy. The result is a paste which can be worked into cavities. As a result of a chemical reaction between the ingredients, the paste sets hard. This material has a silvery appearance and so does not match the teeth.

A modern alternative is a polymer-based material. A polymer involving linear chains would have the problem of being relatively soft, particularly when warm. Thus the polymer needs to have cross-linked chains. To the polymer a white filler ($SiO_2$) is added. This gives a basic white colour which can be modified by the dentist by adding pigments to give a colour match with the teeth. The polymer is a polyurethane to which a hardener is added just before the mixture is used for filling. The hardener cross-links polymer chains to give a hard, rigid product.

### 6.3.9 Optical fibres

Optical fibres are used for the transmission of signals, such as those generated in the telephone system. Properties required of such fibres are that signals can be transmitted over large distances, i.e. the power lost per kilometre is small, and that a pulse input at one end arrives at the far end of the fibre as a similar size pulse that has not become appreciably broadened. The term *dispersion* is used for the spreading out of a pulse.

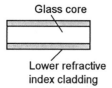

Consider a beam of light shone directly into the end of a glass rod (Figure 6.13). Rays (1) parallel to the axis of the rod will pass straight down the rod and emerge from the far end. Rays (2) incident with the glass–air interface at angles greater than the critical angle will be internally reflected and so pass down the rod and emerge from the far end. The critical angle $C$ is given by (see section 2.5):

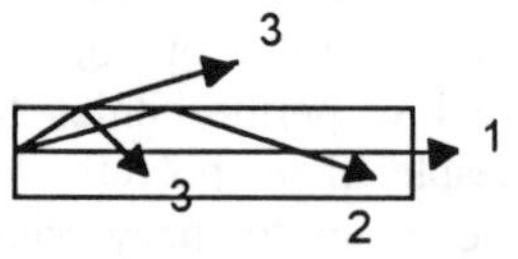

Figure 6.13  *Light rays along a glass rod*

$$n_g \sin C = n_a$$

where $n_g$ is the absolute refractive index of the glass and $n_a$ the absolute refractive index of air. Thus for glass with a refractive index of 1.5, air being effectively 1.0, the critical angle is about 42°.

Some rays (3) will be incident at such an angle with the glass-air interface that they give both a reflected ray and a refracted ray which emerges from the glass. Thus some of the energy of that ray is lost at each reflection. If we increase the critical angle, fewer rays will be refracted out of the rod. This can be done by cladding the glass in a coating of lower refractive index glass (Figure 6.14). Then we have:

Figure 6.14  *Clad glass fibre*

$$n_g \sin C = n_c$$

where $n_c$ is the absolute refractive index of the cladding. If the core has a refractive index greater than that of the cladding by 1% then the critical angle is about 82°.

Thus with such a clad glass fibre, all the rays with incident angles on the core-cladding interface with angles of 82° or more will be internally reflected. The path down the core of such rays will thus be as shown in Figure 6.15. The distance between reflections is:

$$\text{distance between reflections} = d \tan \theta$$

and the distance travelled by the ray of light between reflections is:

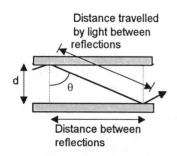

$$\text{distance travelled by the light between reflections} = \frac{d}{\cos \theta}$$

Figure 6.15  *Reflected beam passing down the fibre*

If the core fibre has a diameter of 50 μm then a ray with an incident angle of 82° will follow the path shown in Figure 6.15 and be reflected every $3.56 \times 10^{-4}$ m, i.e. every 0.356 mm. Thus in a length of one metre there will be about 2800 reflections. The distance travelled by the light between reflections is $3.59 \times 10^{-4}$ m. The total distance travelled by the ray of light in passing down a length of 1 m of optical fibre is $2800 \times 3.59 \times 10^{-4}$ m or about 1.009 m. But we also will have rays at angles of 83°, 84°, 85°, ... etc. also reflected down the fibre. Each one of these rays will travel a different distance in passing through 1 m of fibre. For example, for a ray incident at an angle of 85°, the number of reflections made will be 1750, the distance travelled between reflections $5.74 \times 10^{-4}$ m and so the distance travelled in one metre of fibre about 1.004 m. Because these rays travel different distances, the times they take to pass through 1 m of fibre will differ, and spreading out of a pulse occurs, i.e. dispersion. Note that if the critical angle had been smaller, i.e. a glass fibre without cladding had been used, the dispersion would have been greater.

One way of reducing dispersion is to use a fibre which has a core which is only about 5 μm in diameter. Figure 6.16(a) shows the refractive index profile for a 50 μm core in a 125 μm diameter cladding and 6.16(b) that for a 5 μm core in a 125 μm diameter cladding. This 5 μm diameter is about the same order of size as the wavelength of light. Under such circumstances only a single path for transmission is possible. This form of transmission is termed *monomode*.

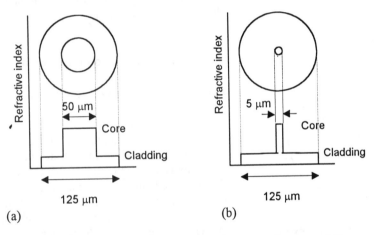

(a)　　　　　　　　　　　　　(b)

Figure 6.16 *Refractive index profiles with step index optical fibres*

An alternative method of reducing dispersion is to use a *graded index fibre*. With such a fibre, the refractive index does not show a sharp step in passing from cladding to core but changes gradually in from the centre of the fibre to the outside (Figure 6.17). The ray which travels straight down the centre of the fibre spends its entire time passing through the highest refractive index glass compared with a ray which is reflected back and forth as it passes down the fibre. That ray spends a significant amount of time in a lower refractive index medium. Because the speed of light is greater in a low refractive index medium than a high refractive medium, though the path lengths differ the times taken to cover them can be made virtually the same. Thus dispersion is reduced.

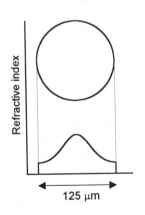

Figure 6.17 *Graded index fibre*

The material used for optical fibres must be one that does not absorb light or cause it to be deviated from its path so that it does not pass through to the end of the fibre. Absorption of light in passing through glass occurs because of impurities in the glass. With silica glass, one part per million of chromium can result in 10% of the light being absorbed in a distance of 10 m. Thus with a long fibre of many kilometres, the presence of even the slightest trace of chromium, or other metallic impurities, in the glass can have a profound effect. Thus impurities need to be removed from the glass. One impurity that is difficult to remove is water. One part per million of water (in fact the $OH^-$ ion) in glass can result in 10% of the light being absorbed in a distance of 1 km. Light can also be scattered from its path through a fibre when it encounters minute fluctuations in the density of the glass. This phenomena is known as *Rayleigh scattering*, the amount of scattering being inversely proportional to the fourth power of the wavelength of the light. If you take a long length of optical fibre and shine white light into one end, the fibre walls are likely to glow blue-green with orange-red light emerging from the other end. This is because the blue-green end of the spectrum is scattered more than the orange-red end. Thus the longer the wavelength of light used for transmission through an optical fibre, the less the loss due to scattering.

### 6.3.10 Magnetic strip on credit cards

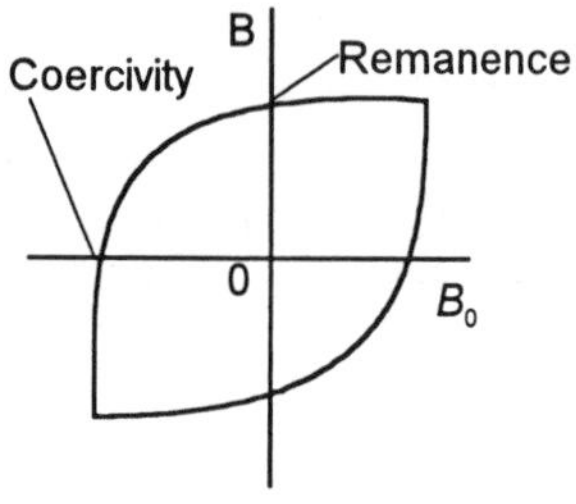

Figure 6.18 *Hysteresis loop*

Credit cards carry magnetic strips that identifies the card-holder by means of a sequence of zeros and ones identified by their magnetism. The magnetic strip is required to retain its magnetism for long periods of time, not easily being demagnetised, and also have a pattern of magnetism which has high enough value to operate the card readers. The material used for the strip is thus required to have a high coercivity, i.e. it is not easily demagnetised, and a high remanence, i.e. it retains a high flux density in the absence of a magnetising field. This would tend to suggest a hysteresis loop which is as near rectangular as possible (Figure 6.18). Such a material is $\gamma Fe_2O_3$. This is a naturally occurring magnetic form of iron oxide. Other materials might have higher coecivities but it is the combination of high coercivity with high remanence which makes this material suitable.

### 6.3.11 Car exhausts

What properties are required of the material used to make exhaust pipes for cars? Such exhaust pipes are externally subject to corrosion as a result of water and solids thrown up from roads, particularly from de-icing salts thrown up in winter. They are also subject to corrosion internally from the corrosive exhaust gases.

Mild steel is not a very good material since corrosion can rapidly result in deterioration of the exhaust. The typical life of such an exhaust was of the order of 18 months, it then having to be replaced. Such exhausts were used up to the 1970s. Aluminium-coated mild steel was able to produce some improvement, the service life extending to typically 30 months. However, steels with a high chromium content, of the order of 12%, were able to considerably improve the length of time for which they could be used, typically giving a service life of the order of four or more years.

A steel containing 0.02% carbon, 0.6% silicon, 0.3% manganese, 11.4% chromium and 0.4% titanium has been developed for car exhausts. The chromium content is responsible for the good corrosion resistance. The titanium gives a steel that can be welded, another requirement for exhaust systems. It also results in a material that can be readily formed to the required shape.

## 6.4 Structures for bridging gaps

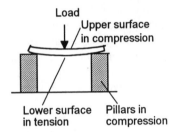

Figure 6.19 *Loads result in bending*

Consider the problems involved in bridging gaps. It could be a bridge across a river or perhaps beams to carry a roof across between two walls.

Suppose we just put a beam of material across the gap (Figure 6.19). What are the forces that the material will have to withstand? The application of loads to the beam will result in bending, with the upper surface of the beam being in compression and the lower surface in tension. The pillars supporting the ends of the beam will be subject to compressive forces. Thus we require materials for the beam that will not break under tensile or compressive forces and, for the supporting pillars, ones which will withstand compressive forces.

Stone is strong in compression and weak in tension. Thus we could use stone for the supporting pillars but a stone beam can present problems. We can use stone only if we keep the tensile forces on the beam down to an acceptable low level. This can be done by having supports close together and using large cross-section stone beams. The greater the cross-section of a beam, the less it curves when subject to a load and so the less its outer surface changes in length. Thus ancient Egyptian and Greek temples tend to have many roof supporting columns relatively short distances apart and very large cross-section beams across their tops. Figure 6.20 illustrates these points.

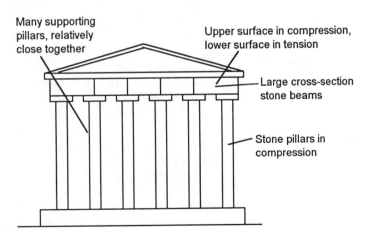

Figure 6.20 *The basic structure when stone beams are used*

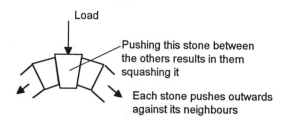

Figure 6.21 *The arch*

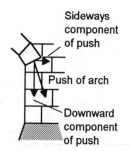

Figure 6.22 *Sideways push of arches*

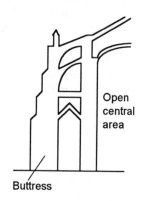

Figure 6.23 *A buttressed cathedral wall*

One way of overcoming the weakness of stone in tension is to build *arches* (Figure 6.21). Each stone in an arch is so shaped that when the force causes a stone to move downwards, the stones on either side resist this and so squash it and put it in compression. Each stone pushes outwards against is neighbours. Thus the net effect of all the downward forces on an arch is to endeavour to straighten it out. The supporting columns must be strong enough to withstand the resulting sideways push of the arch (Figure 6.22) and the foundations of the columns secure enough to withstand the base of the column being displaced.

Cathedrals use arches to span the open central area. The sideways forces due to these arches on the supporting walls is taken account of by providing buttresses. Figure 6.23 illustrates this.

Both stone and brick are strong in compression but weak in tension. Thus arches are widely used when structures are made with such materials. The term *architecture of compression* is often used for such types of structures since they have always to be designed to put the materials into compression.

The end of the eighteenth century saw the introduction into bridge building of a new material, cast iron. Like stone and brick, cast iron is strong in a compression and weak in tension. Thus the iron bridge followed virtually the same form of design as a stone bridge and was in the form of an arch. The world's first iron bridge was built in 1779 over the River Severn. The bridge is about 8 m wide and 100 m long and is still standing.

The introduction of steel, which was strong in tension, enabled the basic design to be changed for bridges and other structures involving the bridging of gaps. We could have the *architecture of tension*. No longer was it necessary to have arches. Instead we could have essentially just a beam resting on two supports. It no longer matters that the lower surface of the beam will be in tension. For a bridge, instead of a continuous beam we can have a hollow beam-type structure built up of separate members. The result is what is termed a *truss bridge*. Figure 6.24 shows one form of such a truss bridge. As with a simple beam, loading results in the upper part of this structure being in compression and the lower part in tension. The diagonal struts are hinged, some of them subject to compressive and some tensile forces.

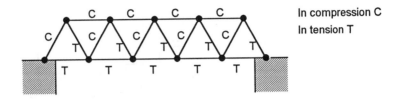

Figure 6.24 *The basic form of a truss bridge*

The *suspension bridge* is a development that is only possible if the materials used are strong in tension. Figure 6.25 shows the basic features of such a bridge. The cable supporting the bridge deck is in tension, and such forms of bridge only became feasible with the development of materials that are strong in tension. The forces acting on the cable have components which pull inwards on the supporting towers. This force means that the cables have to have firm anchorage points.

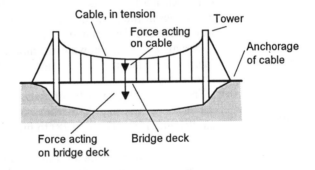

Figure 6.25 *The basic form of a suspension bridge*

Many modern bridges use reinforced and prestressed concrete. This material used the reinforcement to enable the concrete, which is weak in tension but strong in compression, to withstand tensile forces. Such bridges also use the material in the form of an arch in order to keep the material predominantly in compression.

Modern buildings can also use the architecture of tension. Figure 6.26 shows the basic structure of a modern office block. It has a central spine from which cantilevered arms of steel or steel-reinforced concrete stick out. The walls, often just glass in metal frames, are hung between the arms. The cantilevered arms are subject to the loads on a floor of the building and bend, the upper surface being in compression and the lower in tension.

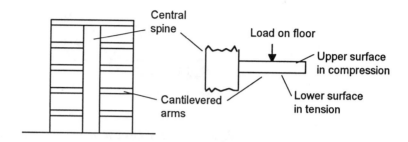

Figure 6.26 *Basic structure of a tower block*

### 6.4.1 Forces and structures

The effect of a force on a structure depends on its magnitude, its direction and the point at which it is applied. Quantities, like force, that require both the magnitude and the direction to be specified before their effects can be determined are called *vectors*. The term *line of action* is used for the line along which a force acts. A vector quantity can be represented by a straight line drawn in the same direction as the vector and with a length which is proportional to the magnitude of the vector. Thus, for example, a force of 20 N acting vertically downwards might be represented by the line shown in Figure 6.27. The arrow on the line indicates the direction in which the force is acting

The resultant effect of two forces with lines of action meeting at a point can be found by using the *parallelogram law*. The procedure for using the parallelogram law (Figure 6.28) to find the resultant force is:

1  Select a suitable scale for drawing lines to represent the forces.
2  Draw an arrowed line to represent the first force.
3  From the start of the first line, draw an arrowed line to represent the second force.
4  Complete the parallelogram by drawing lines parallel to these lines.
5  The result force is represented by the line drawn as the diagonal from the start point.

If we have a single force, we can use the parallelogram law to replace it by two other forces, i.e. we represent the single force by the diagonal and find the two sides of the parallelogram. Any force is thus said to be capable of being resolved into two components. If the two components are at right angles to each other, then we have the situation shown in Figure 6.29. Then we have, for the two components $F_y$ and $F_x$:

$$F_y = F \sin \theta$$

and

$$F_x = F \cos \theta$$

Figure 6.27 *Representation of a force*

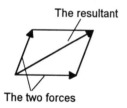

Figure 6.28 *The parallelogram law*

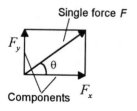

Figure 6.29 *Resolving a force into its components*

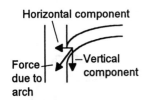

**Figure 6.30** *Sideways force due to an arch*

Thus, for example, if we consider an arch and have a force $F$ acting outwards due to the arch, we can replace this force by a vertically acting component and a horizontally acting component and see what horizontal force has to be balanced in order to stop the arch pushing its side support outwards (Figure 6.30).

Consider the use of a buttress to stop an arch pushing its side support over. The sideways thrust of the arch has a force, the top weight of the buttress, added to it (Figure 6.31(a)) to give a resultant force which is nearer the vertical (Figure 6.31(b)). The heavier the top weight, the more vertical the resultant force, hence the addition of pinnacles and statues. As we progress down the wall, the weight of wall above each point increases. Thus the line of action of the force steadily changes (Figure 6.31(c)), until ideally it becomes vertical at the base of the wall.

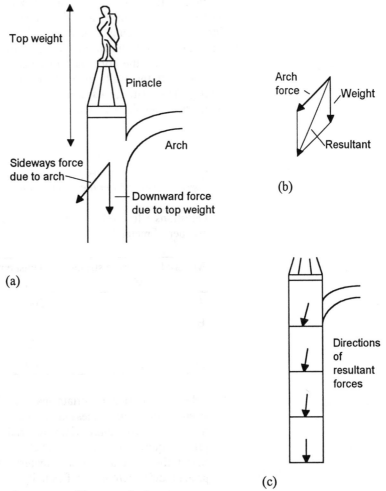

**Figure 6.31** *The use of a buttress*

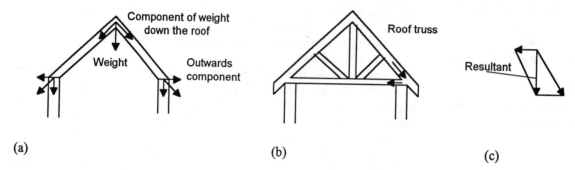

**Figure 6.32** *Supporting a house roof*

While buttresses can be quite decorative, they would not suit or be economic for a house. How then does a house support its roof without a sideways force being produced which would force the walls outwards (Figure 6.32(a))? The answer is a roof truss (Figure 6.32(b)). The weight of the roof causes the horizontal member of the truss to be in tension. Thus the forces at the walls give rise to a resultant which is vertical (Figure 6.32(c)). The material used for roof trusses is generally wood as it is a relatively strong material when subject to tension.

## Problems

*Questions 1 to 16 have four answer options: A, B, C and D. Choose the correct answer from the answer options.*

Questions 1 to 3 relate to the following table, which lists properties of a number of metals.

| Material | Tensile strength MPa | Tensile modulus GPa | Percentage elongation |
|---|---|---|---|
| A | 400 | 200 | 35 |
| B | 900 | 200 | 12 |
| C | 300 | 70 | 15 |
| D | 250 | 100 | 15 |

1  Select the most appropriate material from the list for use where high strength and high stiffness are required.
2  Select the most appropriate material from the list for use where the prime requirement is for high ductility.
3  Select the most appropriate material from the list for use where the prime requirement is high flexibility.

Questions 4 to 7 relate to the following table, which lists properties of a number of metals.

| Material | Thermal conductivity $W\,m^{-1}\,K^{-1}$ | Coefficient of expansion $10^{-6}\,K^{-1}$ | Specific heat capacity $J\,kg^{-1}\,K^{-1}$ |
|---|---|---|---|
| A | 160 | 23 | 880 |
| B | 0.2 | 100 | 200 |
| C | 0.6 | 0.7 | 600 |
| D | 0.1 | 200 | 180 |

**4** Select the most appropriate material from the list for use as a handle to be held by hand for a pan of hot liquid.

**5** Select the most appropriate material from the list for use when the requirement is for a material for use in a heat exchanger where heat is to be conducted away as fast as possible.

**6** Select the most appropriate material from the list for use when the requirement is for a material that will show large increases in temperature when heated.

**7** Select the most appropriate material from the list for use in a product that is subject to temperature fluctuations and which must maintain its dimensional stability.

Questions 8 to 10 relate to the following information about key properties that might determine the selection of a material for a particular use.

A Ductility.
B Toughness.
C Electrical conductivity.
D Thermal conductivity.

**8** Select the key property of relevance in determining the selection of a material for use as the casing for a portable telephone.

**9** Select the key property of relevance in determining the selection of a material for use as the casing for a mains electric plug.

**10** Select the key property of relevance in determining the selection of a material for a product that is to be cold formed.

**11** Decide whether each of these statements is TRUE (T) or FALSE (F).

Figure 6.33 shows a simple girder bridge.
(i) Member AB is in compression.
(ii) Member BC is in tension.

A (i) T  (ii) T
B (i) T  (ii) F
C (i) F  (ii) T
D (i) F  (ii) F

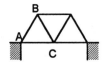

Figure 6.33 *Girder bridge*

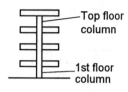

Figure 6.34 *Tower block*

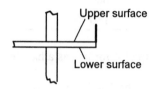

Figure 6.35 *Cantilevered balcony*

**12** A vertical building has floors cantilevered out from a central column (Figure 6.34). If each floor has the same weight $W$ then, if the weight of the column is neglected, the stress in the first floor column will be:

A   Four times that in the top floor column.
B   Twice that in the top floor column.
C   The same as that in the top floor column.
D   One quarter that in the top floor column.

**13** Decide whether each of these statements is TRUE (T) or FALSE (F).

Figure 6.35 shows a cantilevered balcony.
(i) The upper surface will be in compression.
(ii) The lower surface will be in tension.

A   (i) T   (ii) T
B   (i) T   (ii) F
C   (i) F   (ii) T
D   (i) F   (ii) F

**14** Decide whether each of these statements is TRUE (T) or FALSE (F).

The stone beams joining the tops of columns in ancient Greek temples are short and thick because:
(i) This reduces the tensile forces in the lower surfaces of the beam.
(ii) Stone is weak in tension.

A   (i) T   (ii) T
B   (i) T   (ii) F
C   (i) F   (ii) T
D   (i) F   (ii) F

**15** Decide whether each of these statements is TRUE (T) or FALSE (F).

Buttresses are used with arches in cathedrals in order to:
(i) Change the line of action of the forces due to the arch to the vertical.
(ii) Reduce the horizontal force acting on the supporting walls.

A   (i) T   (ii) T
B   (i) T   (ii) F
C   (i) F   (ii) T
D   (i) F   (ii) F

**16** Figure 6.36 shows an arch. The force at the point P due to the load acting on the arch is in the direction

A   →
B   ←
C   ↙
D   ↗

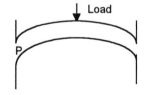

Figure 6.36 *A loaded arch*

17 Make reasoned proposals for materials for the following products:
  (a) Domestic window catches.
  (b) Structural I-beams for use in building construction.
  (c) Rainwater gutters and drainpipes.
  (d) A domestic washing-up bowl.
  (e) A pipe through which sea water can be pumped.
  (f) A small fan in a vacuum cleaner.
  (g) The lenses for the rear lights of cars.
  (h) A camshaft for a car.
  (i) The casing for a hand-held power tool.
  (j) The blades for a Flymo hover-mower.

18 Investigate the materials used with the following products and give reasons why they might have been chosen in preference to others:
  (a) The casing for mains electric plugs.
  (b) Spades.
  (c) Domestic cold and hot water pipes.
  (d) The casing for the body of a vacuum cleaner.
  (e) Joists to support floors in a small house.

# Appendix: Materials

This appendix lists commonly encountered metals, polymers, ceramics, composites and semiconductors, and their characteristics and uses.

**Metals**
**Aluminium** Used in commercially pure form and alloyed with copper, manganese, silicon, magnesium, tin and zinc. Some alloys can be heat treated. Aluminium and its alloys have a low density, high electrical and thermal conductivity and excellent corrosion resistance. Tensile strength tends to be of the order of 150 to 400 MPa with the tensile modulus about 70 GPa. There is a high strength-to-weight ratio. They are used for such applications as engine parts, car trims, aircraft structures, fan blades, typewriter frames, cooking utensils, storage tanks, pressure vessels and chemical equipment.

**Chromium** Chromium is mainly used as an alloying element in stainless steels, heat-resistant alloys and high strength alloy steels. It is generally used for the corrosion and oxidation resistance it confers on the alloys.

**Cobalt** Cobalt is widely used as an alloy for magnets, typically 5 to 35% cobalt with 14 to 30% nickel, and 6 to 13% aluminium. It is also used for alloys that have high strength and hardness at room and high temperatures. These are often referred to as Stellites. Cobalt is also used as an alloying element in steels.

**Copper** Copper is very widely used in the commercially pure form and alloyed in the form of brasses, bronzes, cupro-nickels, and nickel silvers. In the commercially pure form, it is particularly used for electrical conductors. Brasses are copper-zinc alloys containing up to 43% zinc, and are used for decorative and architectural items, coins, medals, fasteners, plumbing pipes, locks, hinges, pins and rivets. Bronzes are copper-tin alloys, and are used for screws, bolts, rivets, springs, clips, bellows and diaphragms. Copper-aluminium alloys are referred to as aluminium bronzes and are used for nuts, bolts, bearings and heat-exchanger tubes. Copper-silicon alloys are termed silicon bronzes and used for chemical and marine plant items. Copper-beryllium alloys are called beryllium bronzes and used for springs, clips and fasteners. Copper-nickel alloys are called cupro-nickels and are used for coins, medals and where high corrosion resistance is required to sea water. If zinc is added to copper-nickel alloys, the resulting alloy is termed a nickel-silver. It is used for clock and watch components, rivets, screws, clips and decorative items. Copper and its alloys have good corrosion resistance, high electrical and thermal conductivity, good machine-ability, can be joined by soldering, brazing and welding, and generally have good properties at low temperatures. The alloys have tensile strengths ranging from about 180 to 300 MPa and a tensile modulus about 20 to 28 GPa.

**Germanium** This is a semiconductor and used, with doping, for the manufacture of semiconductor devices.

**Gold** Gold is very ductile and readily cold worked. It has good electrical and thermal conductivity.

**Iron** The term ferrous alloys is used for the alloys of iron. These include carbon steels, cast irons, alloy steels and stainless steels. Steels have 0.05 to 2% carbon, cast irons 2 to 4.3% carbon. The term carbon steel is used for those steels in which essentially just iron and carbon are present. Steels with between 0.10 and 0.25% are termed mild steels, between 0.20 and 0.50% medium-carbon steels and 0.50 to 2% carbon as high-carbon steels. With such steels in the annealed state, the tensile strength increases from about 250 MPa at low carbon content to 900 MPa at high carbon content, the higher the carbon content the more brittle the alloy. Mild steels are general-purpose steels and used for such applications as joists in buildings, bodywork for cars and ships, screws, nails and wire. Medium-carbon steels are used for shafts and parts in car transmissions, suspensions and steering. High-carbon steels are used for machine tools, saws, axes, hammers, cold chisels, punches and drills. The term low alloy steel is used for alloy steels when the alloying additions are less than 2%, medium alloy between 2 and 10% and high alloy when over 10%. In all cases the carbon content is less that 1%. Example of low alloy steels are manganese steels with strengths of the order of 500 MPa in the annealed state and 700 MPa when quenched and tempered. They are used in applications where higher strengths are required than are possible with carbon steels, e.g. axles and shafts. Stainless steels are high alloy steels with more than 12% chromium. They are used where high corrosion is required, e.g. chemical and food-processing equipment. The modulus of elasticity of steels tend to be about 200 to 207 GPa. Cast irons typically have tensile strengths of the order of 150 to 500 MPa, a tensile modulus of about 100 to 170 GPa, and percentage elongations that are often fairly low, a few per cent. They can be hard and brittle. They are used for machine castings, manhole covers, heavy-duty piping and machine tool beds.

**Lead** Other than its use in lead storage batteries, it finds a use in lead–tin alloys as a metal solder and in steels to improve the machinability.

**Magnesium** Magnesium is used alloyed mainly with aluminium, zinc and manganese. The alloys have a very low density and though tensile strengths are only of the order of 250 MPa, there is a high strength-to-weight ratio. The alloys have a low tensile modulus, about 40 GPa. They have good machinability. They find uses in such applications as instrument casings, power tool and electric motor components, car wheels and in the aircraft industry for parts where weight is an important consideration.

**Molybdenum** Molybdenum has a high density, high electrical and thermal conductivity and low thermal expansivity. At high temperatures, it oxides. It is used for electrodes and support members in electronic tubes and light bulbs, and heating elements for furnaces. Molybdenum is, however, more widely used as an alloying element in steels. In tool steels it improves hardness, in stainless steels it improves corrosion resistance, and in general, in steels it improves strength, toughness and wear resistance.

**Nickel** Nickel is used for a number of alloys with excellent corrosion resistance and strength at high temperatures. The alloys are basically nickel–copper and nickel–chromium–iron and have tensile strengths between about 350 and 1400 MPa and a tensile modulus about 220 GPa. They are used for pipes and containers in the chemical industry where high resistance to corrosive atmospheres is required, food-processing equipment, and applications, such as gas turbine blades and parts, where strength at high temperatures is required.

**Niobium** It has a high melting point, good oxidation resistance and low modulus of elasticity. Niobium alloys are used for high temperature items in turbines and missiles. It is used as an alloying element in steels.

**Palladium** This metal is highly resistant to corrosion. It is alloyed with gold, silver or copper to give metals that are used mainly for electrical contacts.

**Platinum** The metal has a high resistance to corrosion, is very ductile and malleable, but expensive. It is widely used in jewellery. Alloyed with elements such as iridium and rhodium, the metal is used in instruments for items requiring a high resistance to corrosion.

**Silicon** This is a semiconductor and is used, with doping, for the manufacture of semiconductor devices.

**Silver** Silver has a high thermal and electrical conductivity, and is very soft and ductile.

**Tantalum** Tantalum is a high melting point, highly acid-resistant, very ductile metal. Tantalum–tungsten alloys have high melting points, high corrosion resistance and high tensile strength.

**Tin** Tin has a low tensile strength, is fairly soft and can be very easily cut. Tin plate is steel plate coated with tin, the tin conferring good corrosion resistance. Solders are essentially tin alloyed with lead and sometimes antimony. Tin alloyed with copper and antimony gives a material widely used for bearing surfaces.

**Titanium** Titanium as a commercially pure metal or alloy has a high strength coupled with a relatively low density. It retains its properties over a wide temperature range and has excellent corrosion resistance. Tensile strengths are typically of the order of 1100 MPa and tensile modulus about 110 GPa. Titanium and its alloys are used for jet engine parts and for marine and chemical plant parts.

**Tungsten** This is a dense metal with the highest melting point of any metal (3410°C). It is used for light bulb and electronic tube filaments, electrical contacts and as an alloying elements in steels.

**Zinc** Zinc has very good corrosion resistance and hence finds a use as a coating for steel, the product being called galvanised steel. It has a low melting point and hence zinc alloys are used for products such as small toys, cogs, shafts, door handles, etc., produced by die casting. Zinc alloys are generally about 96% zinc with 4% aluminium and small amounts of other elements or 95% zinc with 4% aluminium, 1% copper and small amount of other elements. Such alloys have tensile strength of about 300 MPa, elongations of about 7 to10% and hardness of about 90 BH.

**Polymers**     **Acrylonitrile butadiene styrene (ABS)** This is a thermoplastic polymer giving a range of opaque materials with good impact resistance, ductility and moderate tensile (17 to 58 MPa) and compressive strength. It has a reasonable tensile modulus (1.4 to 3.1 GPa) and hence stiffness, with good chemical resistance. It is used as casings for telephones, vacuum cleaners, hair dryers, TV sets and radios.

**Acetals** Acetals, i.e. polyacetals, are thermoplastics with properties and applications similar to those of nylons. A high tensile strength (70 MPa) is retained in a wide range of environments. They have a high tensile modulus (3.6 GPa) and hence stiffness, high impact resistance and a low coefficient of friction. Ultraviolet radiation causes surface damage. They are used as pipe fittings, parts for water pumps and washing machines, car instrument housings, bearings and hinges.

**Acrylics** Acrylics are transparent thermoplastics, trade names for such materials including Perspex and Plexiglass. They have high tensile strength (50 to 70 MPa) and tensile modulus (2.7 to 3.5 GPa), hence stiffness, good impact resistance and chemical resistance, but a large thermal expansivity. They are used for light fittings, lenses for car lights, signs and shower cabinets.

**Butadiene acrylonitrile** This is an elastomer, generally referred to as nitrile or Buna-N rubber (NBR). It has excellent resistance to fuels and oils and is used for gaskets, hoses, seals and rollers.

**Butadiene styrene** This is an elastomer and is very widely used as a replacement for natural rubber because of its cheapness. It has good wear and weather resistance, good tensile strength, but poor resilience, poor fatigue strength and low resistance to fuels and oils. It is used in the manufacture of car tyres, hose pipes and conveyor belts.

**Butyl** Butyl, i.e. isobutene–isoprene copolymer, is an elastomer. It is extremely impermeable to gases, and it is used for the inner linings of tubeless tyres, steam hoses and diaphragms.

**Cellulosics** This term encompasses cellulose acetate, cellulose acetate butyrate, cellulose acetate propionate, cellulose nitrate and ethyl cellulose. All are thermoplastics. Cellulose acetate is a transparent material. Additives are required to improve toughness and heat resistance. Cellulose acetate butyrate is similar to cellulose acetate but less temperature sensitive and with a greater impact strength. It has a tensile strength of 18 to 48 MPa and a tensile modulus of 0.5 to 1.4 GPa. Cellulose nitrate colours and becomes brittle on exposure to sunlight, and it also burns rapidly. Ethyl cellulose is tough and has low flammability. Cellulosics are used for spectacle frames, tool handles, toys, containers, cable insulation and lenses for instrument panel lights.

**Chlorosulphonated polyethylene** This is an elastomer, trade name Hypalon. It has excellent resistance to ozone with good chemical resistance, fatigue and impact properties. It is used for flexible hose for oil and chemicals, tank linings, cable insulation and shoe soles.

**Epoxies** Epoxy resins are, when cured, thermosets. They are frequently used with glass fibres to form composites or laminates. Such composites have high strength, of the order of 200 to 420 MPa, and stiffness, about 21 to 25 GPa.

**Ethylene propylene** This is an elastomer with very high resistance to oxygen, ozone and heat. It is used for electrical insulation, hoses and belts.

**Ethylene vinyl acetate** This is an elastomer that has good flexibility, impact strength and electrical insulation properties. It is used for cable insulation, flexible tubing and gaskets.

**Fluorocarbons** These are polymers consisting of fluorine attached to carbon chains. See *polytetrafluoroethylene*.

**Fluorosilicones** See *silicone rubbers*.

**Melamine formaldehyde** The resin, a thermoset, is widely used for impregnating paper to form decorative panels, and as a laminate for table and kitchen unit surfaces. It is also used with fillers for moulding knobs, handles, cups, saucers, toys and light fittings. It has good chemical and water resistance, good colourability and good mechanical strength (55 to 85 MPa) and stiffness (7.0 to 10.5 GPa).

**Natural rubber** This is an elastomer. It is inferior to synthetic rubbers in oil and solvent resistance and oxidation resistance. It is attacked by ozone. It is used for tyres, hose and gaskets.

**Nylons** The term nylon is used for a range of thermoplastic materials having the chemical name of polyamides. A numbering system is used to distinguish between the various forms, the most common engineering ones being nylon 6, nylon 6.6 and nylon 11. Nylons are translucent materials with high tensile strength and of medium stiffness. Tensile strengths are typically about 75 MPa and the tensile modulus about 1.1 to 3.3 GPa. Additives such as glass fibres are used to increase strength. Nylons have low coefficients of friction, which can be further reduced by suitable additives. For this reason they are widely used for gears, rollers, bearings and bushes. They are also used for housings for power tools, electric plugs and sockets and as fibres in clothing. All nylons absorb water.

**Phenol formaldehyde** This is a thermoset, known as Bakelite, and is mainly used with reinforcement. It is low cost and has good heat resistance, dimensional stability and water resistance. Unfilled it has a tensile strength of 35 to 55 MPa and a tensile modulus of 5.2 to 7.0 GPa. It is used for electrical plugs and sockets, switches, door knobs and handles.

**Polyacetal** See *acetals*.

**Polyamides** See *nylons*.

**Polycarbonates** Polycarbonates are transparent thermoplastics with high impact strength, high tensile strength (55 to 65 MPa), high dimensional stability and good chemical resistance. They are moderately stiff (2.1 to 2.4 GPa), have good heat resistance and can be used at temperatures up to 120°C. They are used for street lamp covers, infant feeding bottles, machine housings, safety helmets, cups and saucers.

**Polychloroprene** This, usually called neoprene, is an elastomer. It has good resistance to oils and good weathering resistance. It is used for oil and petrol hoses, gaskets, seals, diaphragms and chemical tank linings.

**Polyesters** Two forms are possible, thermoplastics and thermosets. Thermoplastic polyesters have good dimensional stability, excellent electrical resistivity and are tough. They discolour when subject to ultraviolet radiation. Thermoset polyesters are generally used with glass fibres to form composite materials that are used for boat hulls, building panels and stackable chairs.

**Polyethylene** Polyethylene, or polythene, is a thermoplastic material. There are two main types: low density (LDPE), which has a branched polymer chain structure, and high density (HDPE), with linear chains. Materials composed of blends of the two forms are available. LDPE has a fairly low tensile strength (8 to 16 MPa) and tensile modulus (0.1 to 0.3 GPa), with HDPE being stronger (22 to 38 MPa) and stiffer (0.4 to 1.3 GPa). Both forms have good impermeability to gases and very low absorption rates for water. LDPE is used for bags, squeeze bottles, ball-point pen tubing, and wire and cable insulation. HDPE is used for piping, toys and household ware.

**Polyethylene terephthalate (PET)** This is a thermoplastic polyester. It has good strength (50 to 70 MPa) and stiffness (2.1 to 4.4 GPa), is transparent and has good impermeability to gases. It is widely used as bottles for fizzy drinks. It is also used for electrical plugs and sockets, recording tape and wire insulation.

**Polypropylene** Polypropylene is a thermoplastic material with a low density, reasonable tensile strength (30 to 40 MPa) and stiffness (1.1 to 1.6 GPa). Its properties are similar to those of polyethylene. Additives are used to modify the properties. It is used for crates, containers, fans, car fascia panels, radio and TV cabinets, toys and chair shells.

**Polypropylene oxide** This is an elastomer with excellent impact and tear strengths, good resilience and good mechanical properties. It is used for electrical insulation.

**Polystyrene** Polystyrene is a transparent thermoplastic. It has moderate tensile strength (35 to 60 MPa), reasonable stiffness (2.5 to 4.1 GPa), but is fairly brittle and exposure to sunlight results in yellowing. It is attacked by many solvents. Toughened grades, produced by blending with rubber, have better impact properties. They have a strength of about 17 to 42 MPa and stiffness of 1.8 to 3.1 GPa. This form is used as vending machine cups, casings for cameras, radios and TV sets. Foamed, or expanded as it is generally termed, polystyrene is used for insulation and packaging.

**Polysulphide** This is an elastomer with excellent resistance to oils and solvents, and low permeability to gases. It can, however, be attacked by microorganisms. It is used for cable covering, coated fabrics and sealants in building work.

**Polysulphone** This is a strong, comparatively stiff, thermoplastic which can be used to a comparatively high temperature. It has good dimensional stability and low creep. It has a strength of about 70 MPa and a stiffness of about 2.5 GPa. It burns with difficulty and does not present a smoke hazard. It is used in aircraft as parts on passenger service units, circuit boards and cooker control knobs.

**Polytetrafluoroethylene (PTFE)** PTFE is a tough and flexible thermoplastic that can be used over a very wide temperature range. Because other materials will not bond with it, the material is used as a coating to items where non-stick facilities are required, e.g. non-stick domestic cooking pans.

**Polyvinyls** Polyvinyls are thermoplastics and include polyvinyl acetate, polyvinyl butyl, polyvinyl chloride (PVC), chlorinated polyvinyl chloride and vinyl copolymers. Polyvinyl acetate (PVA) is widely used in adhesives and paints. Polyvinyl butyl (PVB) is mainly used as a coating material or adhesive. PVC has high strength (52 to 58 MPa) and stiffness (2.4 to 3.1 GPa), being a rigid material. It is frequently combined with plasticisers to give a lower strength, less rigid material. Without plasticiser it is used as piping for waste and soil drainage systems, rain water pipes, lighting fittings and curtain rails. With plasticiser it is used for plastic raincoats, bottles, shoe soles, garden hose pipes and inflatable toys. Chlorinated PVC is hard and rigid with excellent chemical and heat resistance. Vinyl copolymers can give a range of properties according to the constituents and their ratio. A common copolymer is vinyl chloride with vinyl acetate in the ratio 85 to 15. This is a rigid material. A more flexible form has the ratio 95 to 5.

**Silicone rubbers** Silicone rubbers or, as they are frequently called, fluorosilicone rubbers, have good resistance to oils, fuels and solvents at high and low temperatures. They do, however, have poor abrasion resistance. They are used for electric insulation seals and shock mounts.

**Styrene–butadiene–styrene** This is called a thermoplastic rubber. Its properties are controlled by the ratio of styrene to butadiene. The properties are comparable to those of natural rubber. It is used for footwear, carpet backing and in adhesives.

**Urea formaldehyde** This is a thermosetting material and has similar applications to melamine formaldehyde. Surface hardness is very good. The resin is also used as an adhesive.

## Ceramics

**Alumina** Alumina, i.e. aluminium oxide, is a ceramic that finds a wide variety of uses. It has excellent electrical insulation properties and resistance to hostile environments. It has a tensile modulus of about 380 GPa. Combined with silica, it is used as refractory bricks.

**Boron** Boron fibres are used as reinforcement in composites with materials such as nickel.

**Boron nitride** This ceramic is used as an electric insulator.

**Carbides** A major use of ceramics is, when bonded with a metal binder to form a composite material, as cemented tips for tools. These are generally referred to as bonded carbides, the ceramics used being generally carbides of chromium, tantalum, titanium and tungsten.

**Cement** There are several types of cement. Typically, Portland cement consists of 60 to 64% calcium oxide, 19 to 25% silicon oxide, 5 to 9% aluminium oxide, 2 to 4% iron oxide. When water is mixed with cement, a reaction occurs which results in a silicate structure being formed. Portland cement is used as the binder/matrix in the manufacture of concrete, a composite material involving cement, gravel and sand.

**Chromium carbide** See *carbides*.

**Chromium oxide** This ceramic is used as a wear-resistant coating.

**Clay products** Many ceramic materials are made primarily from clay to which other materials such as silica and feldspar have been added. The materials are mixed with water and the product formed. It is then dried and fired to produce the ceramic bonds. Earthenware, such as drain pipes, are fired at a low temperature and have a relatively porous structure. China and porcelain have high firing temperatures. This results in some of the mixture being converted into a clear glass and so a translucent material.

**Diamond** Diamond has the largest values of tensile modulus (1200 GPa) and thermal conductivity (3000 W m$^{-1}$ K$^{-1}$). It is thus very hard and feels cold to touch, since heat is very rapidly conducted away from the touching finger.

**Glasses** The basic ingredient of most glasses is silica, a ceramic. Glasses tend to have low ductility, a tensile strength that is markedly affected by microscopic defects and surface scratches, low thermal expansivity and conductivity ( and hence poor resistance to thermal shock), good resistance to chemicals and good electrical insulation properties. Glasses tend to have tensile modulus values of about 70 GPa. Glass fibres are frequently used in composites with polymeric materials.

**Kaolinite** This ceramic is a mixture of aluminium and silicon oxides, being a clay. Large electrical insulators, e.g. those used with overhead high voltage cables, are made by firing a mixture of kaolinite, feldspar and silica.

**Magnesia** Magnesia, i.e. magnesium oxide, is a ceramic and is used to produce a brick called a dolomite refractory.

**Pyrex** This is a heat-resistant glass, being made with silica, limestone and boric oxide. See *glasses*.

**Refractory materials** These are generally composed of mixtures of oxides such as those of silicon, aluminium, magnesium, iron and chromium. They have to withstand high stresses at high temperatures.

**Silica** Silica forms the basis of a large variety of ceramics. For example, it is combined with alumina to form refractory bricks and with magnesium to form asbestos. It is the basis of most glasses.

**Silicon nitride** This ceramic is widely used as fibre reinforcement in materials, such as epoxies. It has a tensile modulus of about 160 GPa.

**Soda glass** This is the common window glass, being made from a mixture of silica, limestone and soda ash. See *glasses*.

**Tantalum carbide** See *carbides*.

**Titanium carbide** See *carbides*.

**Tungsten carbide** See *carbides*.

**Composites**      **Ceramic matrix** Typical ceramic matrix materials are alumina with silicon carbide whiskers (25% silicon carbide: elastic modulus 340 GPa and strength 900 MPa) and Pyrex glass with aluminium oxide fibres (40% fibres: strength about 300 MPa). There is a marked improvement in the toughness of the material when compared with the ceramic matrix material alone. Industrial cutting tools are made with alumina reinforced with silicon carbide whiskers.

**Metal matrix** Alumina and silicon carbide have been used as reinforcement for metals. An example is titanium with silicon carbide fibres (35% silicon carbide: tensile modulus 210 GPa and tensile strength 1700 MPa). The composite is stiffer and stronger than the metal alone, but less ductile.

**Polymer matrix** Typical polymer matrix materials are epoxy with glass fibre (70% glass fibres: tensile modulus 40 GPa and tensile strength 750 MPa in the direction of the fibres), polyester with glass fibre (50%

glass fibre: tensile modulus 40 GPa and tensile strength 760 MPa in the direction of the fibres), nylon with carbon fibre (40% carbon fibre: tensile modulus 20 GPa and tensile strength 250 MPa). Such materials are much stiffer and stronger than the polymer alone and have stiffnesses and strengths as good as some metal alloys.

**Natural composites** Wood is an example of a natural composite. It consists of longitudinal cellulose cells bound together with lignin. The properties along the fibre direction are different from those at right angles to it. For example, the tensile modulus of ash is 16 GPa in the grain direction, 0.9 GPa at right angles to it. Douglas fir is 16 GPa in the grain direction and 0.8 GPa at right angles to it. Bone is another example of a natural composite. Bone, such as the human femur, has a tensile modulus of about 20 GPa and a compressive modulus of about 1 GPa. The compressive strength is about 5 to 20 MPa.

**Laminates** These are made by stacking together sheets of materials that generally have unidirectional properties. When stacked cross-ply, i.e. the fibre orientations in the sheets at right angles to each other, the result can be a product that has the same properties in the two right-angled directions. An example of such a material is plywood with thin layers, plies, of wood being laminated together with an adhesive. Three-ply wood consists of a first layer with fibres in one direction, a second layer with fibres at right angles, and then a final layer with fibres in the same direction as the initial layer. Using plies with a tensile modulus of 16 GPa parallel to the fibres and 1.1 GPa at right angles, three-ply has tensile modulus of 12 GPa parallel to the fibres in the outer layers and 0.9 GPa at right angles. With nine-plies this becomes about 11 GPa and 3 GPa.

## Semiconductors

Note that in the following, the size of the energy gap between the valence and conduction bands is given in units of electron volts (eV). 1 eV is the energy needed to move one electron through a potential difference of 1 V and is thus $1.6 \times 10^{-19}$ J. Diamond, an insulator, has an energy gap of 5 eV.

**Aluminium compounds** The compounds formed between aluminium and arsenic AlAs (energy gap 2.16 eV) and between aluminium and antimony AlSb (energy gap 1.58 eV) are semiconductors. They can be doped to give n- or p-type semiconductors.

**Cadmium compounds** The compounds formed between cadmium and sulphur CdS (energy gap 2.42 eV), between cadmium and tellurium CdTe (energy gap 1.56 eV) and between cadmium and selenium CdSe (energy gap 1.7 eV) are semiconductors. They can be doped to give n- or p-type semiconductors. Cadmium sulphide is doped with gallium, iodine or fluorine to give n-type and with lithium or sodium to give p-type. Indium, aluminium and chlorine are used with CdTe to give n-type, and phosphorus, lithium or sodium to give p-type.

**Gallium compounds** Compounds formed between gallium and arsenic GaAs (energy gap 1.42 eV), between gallium and phosphorus GaP (energy gap 2.26 eV) and between gallium and antimony GaSb (energy gap 0.72 eV) are semiconductors. They can be doped to give n- or p-type semiconductors. For example, tellurium, sulphur, tin, silicon and germanium can be used with GaAs to give n-type, and zinc, chromium, silicon and germanium to give p-type.

**Germanium** This element has an energy gap of 0.66 eV. Phosphorus, arsenic and antimony are used to give n-type semiconductors, and boron or aluminium to give p-type.

**Indium compounds** Compounds formed between indium and arsenic InAs (energy gap 2.16 eV), between indium and phosphorus InP (energy gap 1.35 eV) and between indium and antimony InSb (energy gap 1.58 eV) are semiconductors. They can be doped to give n- or p-type semiconductors.

**Lead compounds** Compounds formed between lead and sulphur PbS (energy gap 0.41 eV), between lead and tellurium (energy gap 0.31 eV) and between lead and selenium PbSe (energy gap 0.27 eV) are semiconductors. They can be doped to give n- or p-type semiconductors.

**Silicon** This element has an energy gap of 1.12 eV. Phosphorus, arsenic and antimony are used to give n-type semiconductors, and boron or aluminium to give p-type.

**Zinc compounds** Compounds formed between zinc and sulphur ZnS (energy gap 3.68 eV), between zinc and tellurium (energy gap 2.2 eV) and between zinc and selenium ZnSe (energy gap 2.7 eV) are semiconductors. They can be doped to give n- or p-type semiconductors.

# Answers

The following are the numerical answers to problems and brief clues as to the form of the answers for other problems.

**Chapter 1**

| 1 C | 2 B | 3 A | 4 A | 5 D | 6 C |
|-----|-----|-----|-----|-----|-----|
| 7 D | 8 B | 9 C | 10 A | 11 C | |

12 Includes: (a) stiff, able to carry weight of table without buckling, can be made in the required shape, (b) stiff, no significant deflection under own weight, hence low density, surface capable of being accurately shaped, (c) low stiffness so easily stretched, not easily permanently deformed, (d) low thermal conductivity, low heat capacity, (e) low density, easily formed, not easily permanently deformed.

**Chapter 2**

| 1 B | 2 C | 3 B | 4 A | 5 A | 6 B |
|-----|-----|-----|-----|-----|-----|
| 7 A | 8 A | 9 C | 10 A | 11 D | 12 B |
| 13 C | 14 D | 15 A | 16 A | 17 B | 18 B |
| 19 C | 20 A | 21 A | 22 D | | |

23 These might include (a) ease of forming in one piece, easily cleaned, stain resistance, waterproof, (b) stiff, strong, cheap, (c) leak proof, suitable for hot liquids, cheap, not easily broken, (d) good conductor, flexible, (e) cheap to make, wear resistant during handling, stiff, (f) withstands changing forces, stiff, strong, withstands impact forces, (g) attractive appearance, cheap to form.

24 (a) Stainless steel, (b) wood, (c) china (a ceramic), (d) copper, (e) alloys of copper (cupro-nickel or bronze depending on the colour of the coins), (f) steel, (g) plastic, e.g. ABS.

25 (a) Modulus, (b) ductility, percentage elongation, (c) fracture toughness, (d) strength, (e) electrical resistivity/conductivity, (f) thermal conductivity, (g) corrosive properties.

26 Strong and brittle.

27 Strong and tough.

28 20 MPa

29 0.67%

20 50 kN

31 12%

32 50 kN

33 The bronze is stronger and more ductile.

34 Stronger in compression, brittle.

35 Very low resistivity, of the order of $10^{-8}\,\Omega\,m$.

36 $0.0125\,\Omega$

37 Brittle, must not be subject to sudden forces or changes in temperature.

38 128 MPa/Mg m$^{-3}$, 33 MPa/Mg m$^{-3}$, 0.78 £/MPa, 6 £/MPa

**Chapter 3**
1 A    2 B    3 C    4 C    5 B    6 D
7 C    8 A    9 C    10 D

11 ±0.05 units
12 (a) 800 kN, (b) ±0.07 kN
13 (a) 51.3 Ω, (b) ±0.07 Ω
14 (a) 39.0 V, (b) ±0.11 kV
15 150 ± 10 Ω
16 33.3 ± 7.2 Ω
17 1.77 ± 0.09 × 10$^6$ mm$^3$
18 787 ± 24 kg/m$^3$
19 61 GPa
20 10.8 MPa
21 (a) 660 MPa, (b) 425 MPa, (c) 200 GPa
22 (a) 480 MPa, (b) 167 GPa
23 Becoming more ductile.
24 Becoming more brittle.
25 As the temperature drops becoming more brittle.
26 HV 198
27 HV 275
28 HV 71
30 HB 217
31 3114 J kg$^{-1}$ K$^{-1}$
32 873 J kg$^{-1}$ K$^{-1}$
33 1.39

**Chapter 4**
1 D    2 B    3 C    4 D    5 C    6 C
7 C    8 B    9 B    10 A    11 B    12 B
13 C    14 A    15 B    16 C    17 C

18 (a) Positive ions surrounded by free electrons, (b) long chain molecules with Van der Waals or hydrogen bonds between them, (c) ionic or covalent bonded metal and non-metallic atoms.
19 (a) Positive ions attracted to electrons, which in turn are attracted to positive ions, (b) electron(s) transfer between atoms and hence attraction between positive and negative ions, (c) sharing of electrons by atoms, hence positive part of atom attracted to shared electrons, which in turn are attracted to positive part of neighbouring atom, (d) temporary uneven distribution of electrons results in an atom inducing a similar uneven distribution in a neighbouring atom, hence force of attraction between the neigbouring unlike charged ends of the atoms.
20 At room temperature, metals have free electrons and hence easy conduction by these; insulators no free electrons and hence no conduction; with semiconductors, some electrons have broken free and left holes and so there is movement of electrons and holes.

**Chapter 5**   1 C      2 A      3 A      4 C      5 A      6 D
            7 D      8 A      9 C      10 B     11 A     12 C

13 A mixture of two or more elements, e.g. iron and carbon in steel.
14 Ferrous alloys have iron as the main constituent, a non-ferrous alloy a metal other than iron.
15 More ductile the bigger the grains.
16 Elongated grains give different properties in the directions of the grains compared with at right angles.
17 Grains become elongated and distorted with an increasing number of dislocations. The tensile strength and hardness increases, the ductility decreases.
18 See Figure 5.2.
19 (a) Large grain, few dislocations, (b) small grain, many dislocations.
20 Increase in dislocations and hence an increase in yield strength, tensile strength and hardness but a decrease in ductility.
21 See Figure 5.2.
22 See Figure 5.2. Above the recrystallisation temperature, no work hardening occurs.
23 The greater the crystallinity, the greater the density, melting point and strength.
24 (a) To protect against UV and resist deterioration, (b) to make more flexible, (c) to reduce cost, increase perhaps strength, impact strength, resistivity, or reduce friction.
25 Makes it more rigid.
26 See Figure 4.18 and associated text in previous chapter.
27 36.4 GPa, in direction of fibres.
28 182.2 GPa
29 205 GPa
30 The long fibres give directionality of properties and a greater improvement in strength and modulus than random fibres which give no directionality.

**Chapter 6**   1 B      2 A      3 C      4 D      5 A      6 A
            7 C      8 B      9 C      10 A     11 A     12 A
            13 D     14 A     15 A     16 C

17 (a) Aluminium alloy, (b) carbon steel, about 0.3% carbon, or low alloy steel such as a chromium-molybdenum steel, (c) unplasticised PVC, (d) high density polyethylene, (e) high density polyethylene, (f) nylon or aluminium alloy, (g) acrylic (polymethyl methacrylate), (h) medium carbon steel, about 0.4% carbon, or low alloy steel, (i) polypropylene or aluminium alloy, (j) polymer ABS.
18 (a) Thermoset, insulator, cheap to form, e.g. urea formaldehyde, (b) medium carbon steel or stainless steel, reasonable strength and hardness, (c) copper, ease of bending, corrosion resistance, (d) polymer, tough, reasonably stiff, coloured, cheap to form, e.g. ABS, (e) stiff, strong in bending, cheap, e.g. wood.

# Index